THE MESSENGER LECTURES have taken place annually at Cornell since 1924, when Hiram J. Messenger, a graduate and Professor of Mathematics, gave a sum of money to encourage eminent personalities from anywhere in the world to visit Cornell and talk to the academic community. In establishing the fund for the lectures, Messenger specified that it is "to provide a course or courses of lectures on the evolution of civilization for the special purpose of raising the moral standard of our political, business, and social life."

In November Professor Richard P. Feynman, the distinguished physicist and educator, gave the 1964 lectures.

THE CHARACTER OF PHYSICAL LAW

RICHARD FEYNMAN

THE MIT PRESS
MASSACHUSETTS INSTITUTE OF TECHNOLOGY
CAMBRIDGE, MASSACHUSETTS, AND LONDON, ENGLAND

Twenty First printing, 1994
First MIT Press Paperback Edition, March 1967

First published in 1965 by
The British Broadcasting Corporation

Copyright ©1965 by Richard Feynman

All rights reserved. This book may not be reproduced, in whole or in part, in any form (except by reviewers for the public press), without written permission from the publishers.

ISBN 0 262 06016 7 (hardcover)
ISBN 0 262 56003 8 (paperback)

Library of Congress Catalog Card Number: 67-14527
Printed in the United States of America

Contents

Foreword

The seven chapters which make up this book were lectures presented as the Messenger Lectures at Cornell University in the United States. They were delivered to an audience of students who wished to know in general terms more about 'The Character of Physical Law'. The lectures were not given from a prepared manuscript, but were delivered extempore from a few notes.

The Messenger Lectures have taken place annually at Cornell since 1924, when Hiram J. Messenger, a graduate and professor of Mathematics, gave a sum of money to encourage eminent personalities from anywhere in the world to visit Cornell and talk to the students. In establishing the fund for the lectures Messenger specified that it is 'to provide a course or courses of lectures on the evolution of civilization for the special purpose of raising the moral standard of our political, business and social life'.

In November Professor Richard P. Feynman, the distinguished physicist and educator, was invited to give the 1964 lectures. He was formerly a professor at Cornell and is now Professor of Theoretical Physics at California Institute of Technology. He has recently been made a Foreign Member of the Royal Society, and is noted not only for his contribution to present day understanding of the laws of physics, but also for his ability to bring his subject alive to the non-physicist.

The chapters in this book are reports of talks which were presented to a packed audience from a large stage which allowed Professor Feynman uninhibited expression of speech and movement. He has international prominence as a lecturer, and is known for his exciting platform manner.

Foreword

This book is intended to serve as a guide or memory aid for television viewers who may see the lectures and wish to have a permanent reminder to refer to. Although it is not in any way to be regarded as a textbook, the student of physics in search of a clearer understanding of the laws will be enlightened by many of the arguments.

Richard Feynman is already known to BBC-1 as one of the physicists in Philip Daly's production 'Men at the Heart of Matter' and for his splendid contribution to 'Strangeness minus three', one of the most fascinating programmes on recent scientific discovery in 1964.

The BBC Science and Features Department became interested when it was known that Professor Feynman was to give the Messenger Lectures. The series is being presented in BBC-2, as part of the Further Education Scheme, and continues in the style of the lectures already given by such distinguished men as Bondi on Relativity, Kendrew on Molecular Biology, Morrison on Quantum Mechanics and Porter on Thermodynamics.

What you are about to read is a transcription of the lectures. They have been checked for scientific accuracy by Professor Feynman. My assistant Fiona Holmes and I have assembled the spoken words so that they are now set down in print. We hope that this book will be acceptable to you. To have worked with Richard Feynman has been a rewarding experience, and we trust that viewers and readers will gain much from this project.

Alan Sleath, *Producer BBC Outside Broadcasts Science and Features Department, June 1965*

The BBC is grateful to the Cornell University News Bureau for permission to reproduce Plate 2, and to the California Institute of Technology for permission to reproduce other photographs and drawings used in Lecture 1.

Students who wish to make a more detailed study of Professor Feynman's work will be interested to know that the books referred to by the Provost in his introduction are published by California Institute of Technology and are entitled *The Feynman Lectures in Physics*.

The Provost of Cornell University
DALE R. CORSON
introduces the Messenger Lecturer for 1964

Ladies and gentlemen, it is my privilege to introduce the Messenger Lecturer, Professor Richard P. Feynman of California Institute of Technology.

Professor Feynman is a distinguished theoretical physicist, and has done much to bring order out of the confusion which has marked much of the spectacular development in physics during the postwar period. Among his honours and awards I will mention only the Albert Einstein Award in 1954. This is an award which is made every third year, and which includes a gold medal and a substantial sum of money.

Professor Feynman did his undergraduate work at M.I.T. and his graduate work at Princeton. He worked on the Manhattan Project at Princeton and later at Los Alamos. He was appointed an Assistant Professor at Cornell in 1944, although he did not assume residence until the end of the War. I thought it might be interesting to see what was said about him when he was appointed at Cornell, so I searched the Minutes of our Board of Trustees . . . and there is absolutely no record of his appointment. There are, however, some twenty references to leaves of absence, salary increases, and promotions. One reference interested me especially. On July 31st 1945 the Chairman of the Physics Department wrote to the Dean of the Arts College stating that 'Dr Feynman is an outstanding teacher and investigator, the equal of whom develops infrequently'. The Chairman suggested that an annual salary of three thousand dollars was a bit too low for a distinguished faculty member, and recommended that Professor Feynman's salary be increased nine hundred dollars. The Dean, in an act of unusual generosity, and with complete disregard for the solvency of the University, crossed out the nine hundred dollars and made it an even thousand. You can see that we thought highly of Professor Feynman even then! Feynman took up residence here at the end of 1945, and spent five highly productive years on our Faculty. He left Cornell in 1950 and went to Cal. Tech., where he has been ever since.

Before I let him talk, I want to tell you a little more about him. Three or four years ago he started teaching a beginning physics course at Cal. Tech., and the result has added a new dimension to his fame – his lectures are now published in two volumes and they represent a refreshing approach to the subject.

In the preface of the published lectures there is a picture of Feynman performing happily on the bongo drums. My Cal. Tech. friends tell me he sometimes drops in on the Los Angeles night spots and takes over the work of the drummer; but Professor Feynman tells me that is not so. Another of his specialities is safe cracking. One legend says that he once opened a locked safe in a secret establishment, removed a secret document, and left a note saying 'Guess who?' I could tell you about the time he learned Spanish before he went to give a series of lectures in Brazil, but I won't.

This gives you enough background, I think, so let me say that I am delighted to welcome Professor Feynman back to Cornell. His general topic is 'The Character of Physical Law', and his topic for tonight is 'The Law of Gravitation, an Example of Physical Law'.

1

The Law of Gravitation, an example of Physical Law

It is odd, but on the infrequent occasions when I have been called upon in a formal place to play the bongo drums, the introducer never seems to find it necessary to mention that I also do theoretical physics. I believe that is probably because we respect the arts more than the sciences. The artists of the Renaissance said that man's main concern should be for man, and yet there are other things of interest in the world. Even the artists appreciate sunsets, and the ocean waves, and the march of the stars across the heavens. There is then some reason to talk of other things sometimes. As we look into these things we get an aesthetic pleasure from them directly on observation. There is also a rhythm and a pattern between the phenomena of nature which is not apparent to the eye, but only to the eye of analysis; and it is these rhythms and patterns which we call Physical Laws. What I want to discuss in this series of lectures is the general characteristic of these Physical Laws; that is another level, if you will, of higher generality over the laws themselves. Really what I am considering is nature as seen as a result of detailed analysis, but mainly I wish to speak about only the most overall general qualities of nature.

Now such a topic has a tendency to become too philosophical because it becomes so general, and a person talks in such generalities, that everybody can understand him. It is then considered to be some deep philosophy. I would like to be rather more special, and I would like to be understood in an honest way rather than in a vague way. So in this first lecture I am going to try to give, instead of only the generalities, an example of physical law, so that you have at least one example of the things about which I am speaking generally. In this way I can use this example again and again to give an instance, or to make a reality out of something which will otherwise be too abstract. I have chosen for my special example of physical law the theory of gravitation, the phenomena of gravity. Why I chose gravity I do not know. Actually it was one of the first great laws to be discovered and it has an interesting history. You may say, 'Yes, but then it is old hat, I would like to hear something about a more modern science'. More recent perhaps, but not more modern. Modern science is exactly in the same tradition as the discoveries of the Law of Gravitation. It is only more recent discoveries that we would be talking about. I do not feel at all bad about telling you about the Law of Gravitation because in describing its history and methods, the character of its discovery, its quality, I am being completely modern.

This law has been called 'the greatest generalization achieved by the human mind', and you can guess already from my introduction that I am interested not so much in the human mind as in the marvel of a nature which can obey such an elegant and simple law as this law of gravitation. Therefore our main concentration will not be on how clever we are to have found it all out, but on how clever nature is to pay attention to it.

The Law of Gravitation is that two bodies exert a force upon each other which varies inversely as the square of the distance between them, and varies directly as the product of their masses. Mathematically we can write that great law down in the formula:

$$F = G\frac{mm'}{r^2}$$

some kind of a constant multiplied by the product of the two masses, divided by the square of the distance. Now if I add the remark that a body reacts to a force by accelerating, or by changing its velocity every second to an extent inversely as its mass, or that it changes its velocity more if the mass is lower, inversely as the mass, then I have said everything about the Law of Gravitation that needs to be said. Everything else is a mathematical consequence of those two things. Now I know that you are not all mathematicians, and you cannot immediately see all of the consequences of these two remarks, so what I would like to do here is to tell you briefly of the story of the discovery, what some of the consequences are, what effect this discovery had on the history of science, what kind of mysteries such a law entails, something about the refinements made by Einstein, and possibly the relation to the other laws of physics.

The history of the thing, briefly, is this. The ancients first observed the way the planets seemed to move in the sky and concluded that they all, along with the earth, went around the sun. This discovery was later made independently by Copernicus, after people had forgotten that it had already been made. Now the next question that came up for study was: exactly how do they go around the sun, that is, with exactly what kind of motion? Do they go with the sun as the centre of a circle, or do they go in some other kind of curve? How fast do they move? And so on. This discovery took longer to make. The times after Copernicus were times in which there were great debates about whether the planets in fact went around the sun along with the earth, or whether the earth was at the centre of the universe and so on. Then a man named Tycho Brahe* evolved a way of answering the question. He thought that it might perhaps be a good idea to look very very carefully and to record exactly where the planets appear in the sky, and then the alternative theories might be distinguished from one another. This is the key of modern science and it was the beginning of the true understanding of Nature – this idea to look at the thing, to record the details, and to hope that in the information thus obtained might lie a clue to one or another theoretical interpretation. So Tycho, a rich man who owned an island near

*Tycho Brahe, 1546–1601, Danish astronomer.

Copenhagen, outfitted his island with great brass circles and special observing positions, and recorded night after night the position of the planets. It is only through such hard work that we can find out anything.

When all these data were collected they came into the hands of Kepler,* who then tried to analyse what kind of motion the planets made around the sun. And he did this by a method of trial and error. At one stage he thought he had it; he figured out that they went round the sun in circles with the sun off centre. Then Kepler noticed that one planet, I think it was Mars, was eight minutes of arc off, and he decided this was too big for Tycho Brahe to have made an error, and that this was not the right answer. So because of the precision of the experiments he was able to proceed to another trial and ultimately found out three things.

First, he found that the planets went in ellipses around the sun with the sun as a focus. An ellipse is a curve all artists know about because it is a foreshortened circle. Children also know because someone told them that if you put a ring on a piece of cord, anchored at each end, and then put a pencil in the ring, it will draw an ellipse (fig. 1).

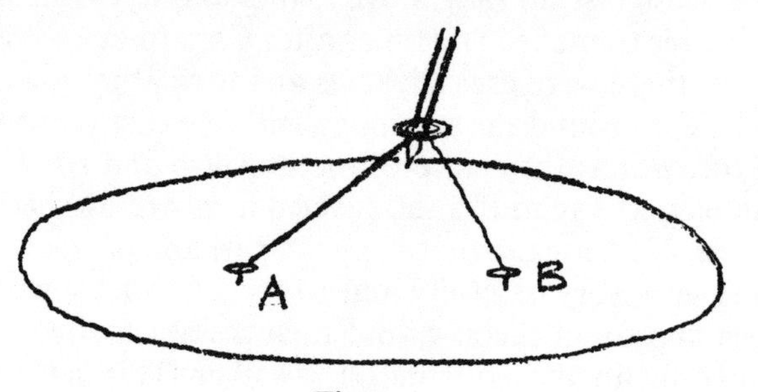

Figure 1

The two points A and B are the foci. The orbit of a planet around the sun is an ellipse with the sun at one focus. The

*Johann Kepler, 1571–1630, German astronomer and mathematician, assistant to Brahe.

next question is: In going around the ellipse, how does the planet go? Does it go faster when it is near the sun? Does it go slower when it is farther from the sun? Kepler found the answer to this too (fig. 2).

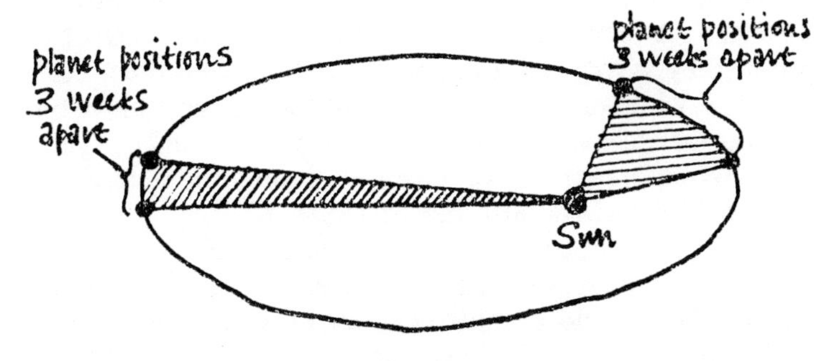

Figure 2

He found that, if you put down the position of a planet at two times, separated by some definite period, let us say three weeks – then in another place on its orbit two positions of the planet again separated by three weeks, and draw lines (technically called radius vectors) from the sun to the planet, then the area that is enclosed in the orbit of the planet and the two lines that are separated by the planet's position three weeks apart is the same, in any part of the orbit. So that the planet has to go faster when it is closer to the sun, and slower when it is farther away, in order to show precisely the same area.

Some several years later Kepler found a third rule, which was not concerned only with the motion of a single planet around the sun but related various planets to each other. It said that the time the planet took to go all around the sun was related to the size of the orbit, and that the times varied as the square root of the cube of the size of the orbit and for this the size of the orbit is the diameter across the biggest distance on the ellipse. Kepler then had these three laws which are summarized by saying that *the orbit forms an ellipse*, and that *equal areas are swept in equal times* and that the *time to go round varies as a three half power of the size*, that is, the square root of the cube of the size. These three laws of Kepler give a complete description of the motion of the planets around the sun.

The next question was – what makes planets go around the sun? At the time of Kepler some people answered this problem by saying that there were angels behind them beating their wings and pushing the planets around an orbit. As you will see, the answer is not very far from the truth. The only difference is that the angels sit in a different direction and their wings push inwards.

In the meantime, Galileo was investigating the laws of motion of ordinary objects at hand on the earth. In studying these laws, and doing a number of experiments to see how balls run down inclined planes, and how pendulums swing, and so on, Galileo discovered a great principle called the principle of inertia, which is this: that if an object has nothing acting on it and is going along at a certain velocity in a straight line it will go at the same velocity in exactly the same straight line for ever. Unbelievable as that may sound to anybody who has tried to make a ball roll for ever, if this idealization were correct, and there were no influences acting, such as the friction of the floor and so on, the ball would go at a uniform speed for ever.

The next point was made by Newton, who discussed the question: 'When it does not go in a straight line *then* what?' And he answered it this way: that a force is needed to change the velocity in any manner. For instance, if you are pushing a ball in the direction that it moves it will speed up. If you find that it changes direction, then the force must have been sideways. The force can be measured by the product of two effects. How much does the velocity change in a small interval of time? That's called the acceleration, and when it is multiplied by the coefficient called the mass of an object, or its inertia coefficient, then that together is the force. One can measure this. For instance, if one has a stone on the end of a string and swings it in a circle over the head, one finds one has to pull, the reason is that although the speed is not changing as it goes round in a circle, it is changing its direction; there must be a perpetually in-pulling force, and this is proportional to the mass. So that if we were to take two different objects, and swing first one and then the other at the same speed around the head, and measure the force in the second one, then that second force is bigger than the other force in proportion as the masses are different. This is a way of measuring the masses by what force is necessary to change the speed. Newton saw from this that, to take a simple example, if a planet is going in a circle around the sun, *no force is needed to make it go sideways, tangentially;* if there were no force at all then it would just keep coasting along. But actually the planet does not keep coasting along, it finds itself later not way out where it would go if there were no force at all, but farther down towards the sun.

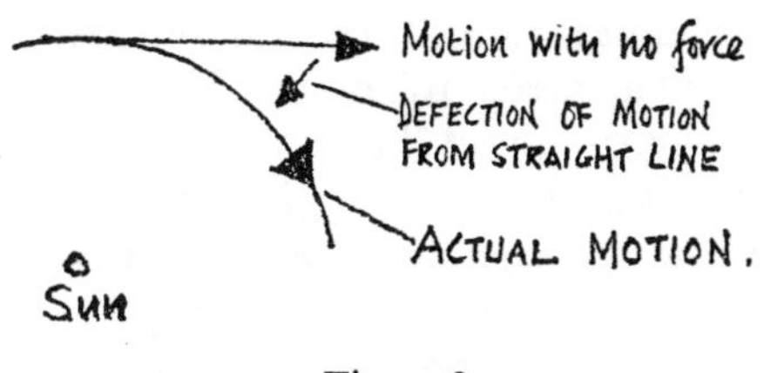

Figure 3

(fig. 3.) In other words, its velocity, its motion, has been deflected towards the sun. So that what the angels have to do is to beat their wings in towards the sun all the time.

But the motion to keep the planet going in a straight line has no known reason. The reason why things coast for ever has never been found out. The law of inertia has no known origin. Although the angels do not exist the continuation of the motion does, but in order to obtain the falling operation we do need a force. It became apparent that the origin of the force was towards the sun. As a matter of fact Newton was able to demonstrate that the statement that equal areas are swept in equal times was a direct consequence of the simple idea that all the changes in velocity are directed exactly towards the sun, even in the elliptical case, and in the next lecture I shall be able to show you how it works, in detail.

From this law Newton confirmed the idea that the force is towards the sun, and from knowing how the periods of the different planets vary with the distance away from the sun, it is possible to determine how that force must weaken at different distances. He was able to determine that the force must vary inversely as the square of the distance.

So far Newton has not said anything, because he has only stated two things which Kepler said in a different language. One is exactly equivalent to the statement that the force is towards the sun, and the other is exactly equivalent to the statement that the force is inversely as the square of the distance.

But people had seen in telescopes Jupiter's satellites going around Jupiter, and it looked like a little solar system, as if the satellites were attracted to Jupiter. The moon is attracted to the earth and goes round the earth and is attracted in the same way. It looks as though everything is attracted to everything else, and so the next statement was to generalize this and to say that every object attracts every object. If so, the earth must be pulling on the moon, just as the sun pulls on the planet. But it is known that the earth is pulling on things – because you are all sitting tightly on your seats in spite of your desire to float into the air. The pull for objects on the earth was well known in the phenomena of gravitation, and it was Newton's idea that maybe the gravitation that held the moon in orbit was the same gravitation that pulled the object towards the earth.

It is easy to figure out how far the moon falls in one second, because you know the size of the orbit, you know the moon takes a month to go around the earth, and if you figure out how far it goes in one second you can figure out how far the circle of the moon's orbit has fallen below the straight line that it would have been in if it did not go the way it does go. This distance is one twentieth of an inch.

The moon is sixty times as far away from the earth's centre as we are; we are 4,000 miles away from the centre, and the moon is 240,000 miles away from the centre, so if the law of inverse square is right, an object at the earth's surface should fall in one second by $\frac{1}{20}$ inch $\times$ 3,600 (the square of 60) because the force in getting out there to the moon, has been weakened by 60 $\times$ 60 by the inverse square law. $\frac{1}{20}$ inch $\times$ 3,600 is about 16 feet, and it was known already from Galileo's measurements that things fall in one second on the earth's surface by 16 feet. So this meant that Newton was on the right track, there was no going back now, because a new fact which was completely independent previously, the period of the moon's orbit and its distance from the earth, was connected to another fact, how long it takes something to fall in one second at the earth's surface. This was a dramatic test that everything is all right.

Further, Newton had a lot of other predictions. He was able to calculate what the shape of the orbit should be if the law were the inverse square, and he found, indeed, that it was an ellipse – so he got three for two as it were. In addition, a number of new phenomena had obvious explanations. One was the tides. The tides were due to the pull of the moon on the earth and its waters. This had sometimes been thought of before, with the difficulty that if it was the pull of the moon on the waters, making the water higher on the side where the moon was, then there would only be one tide a day under the moon (fig. 4), but actually we know

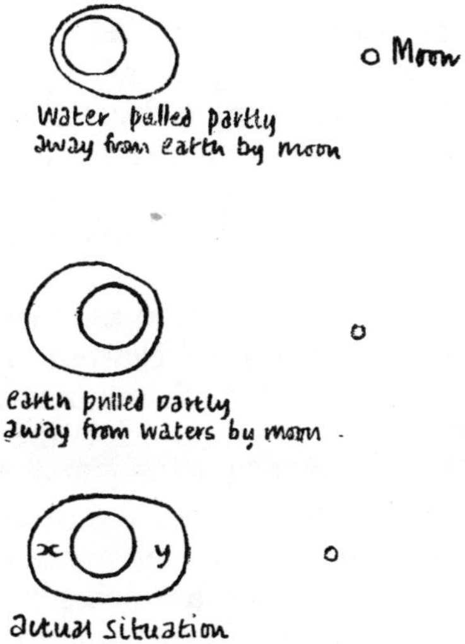

Figure 4

there are tides roughly every twelve hours, and that is two tides a day. There was also another school of thought that came to a different conclusion. Their theory was that it was the earth pulled by the moon away from the water. Newton was actually the first one to realize what was going on; that the force of the moon on the earth and on the water is the same at the same distance, and that the water at y is closer to the moon and the water at x is farther from the moon than the rigid earth. The water is pulled more towards the moon at y, and at x is less towards the moon than the earth, so there is a combination of those two pictures that makes a double tide. Actually the earth does the same trick as the moon, it goes around in a circle. The force of the moon on the earth is balanced, but by what? By the fact that just as the moon goes in a circle to balance the earth's force, the earth is also going in a circle. The centre of the circle is somewhere inside the earth. It is also going in a circle to balance the moon. The two of them go around a common centre so the forces are balanced for the earth, but the water at x is pulled less, and at y more by the moon and it bulges out at both sides. At any rate tides were then explained, and the fact that there were two a day. A lot of other things became clear: how the earth is round because everything gets pulled in, and how it is not round because it is spinning and the outside gets thrown out a little bit, and it balances; how the sun and moon are round, and so on.

As science developed and measurements were made more accurate, the tests of Newton's Law became more stringent, and the first careful tests involved the moons of Jupiter. By accurate observations of the way they went around over long periods of time one could check that everything was according to Newton, and it turned out to be not the case. The moons of Jupiter appeared to get sometimes eight minutes ahead of time and sometimes eight minutes behind time, where the time is the calculated value according to Newton's Laws. It was noticed that they were ahead of schedule when Jupiter was close to the earth and behind schedule when it was far away, a rather odd circumstance. Mr Roemer,* having confidence in the Law of Gravitation, came to the interesting conclusion that it takes light some time to travel from the moons of Jupiter to the earth, and what we are looking at when we see the moons is not how they are now but how they were the time ago it took the light to get here. When Jupiter is near us it takes less time for the light to come, and when Jupiter is farther from us it takes longer time, so Roemer had to correct the observations for the differences in time and by the fact that they were this much early or that much late. In this way he was able to determine the velocity of light. This was the first demonstration that light was not an instantaneously propagating material.

I bring this particular matter to your attention because it illustrates that when a law is right it can be used to find another one. If we have confidence in a law, then if something appears to be wrong it can suggest to us another phenomenon. If we had not known the Law of Gravitation we would have taken much longer to find the speed of light, because we would not have known what to expect of Jupiter's satellites. This process has developed into an avalanche of discoveries, each new discovery permits the tools for much more discovery, and this is the beginning of the avalanche which has gone on now for 400 years in a continuous process, and we are still avalanching along at high speed.

Another problem came up – the planets should not really go in ellipses, because according to Newton's Laws they are not only attracted by the sun but also they pull on each other a little – only a little, but that little is something, and will alter the motion a little bit. Jupiter, Saturn and Uranus were big planets that were known, and calculations were made about how slightly different from the perfect ellipses of Kepler the planets ought to be going by the pull of each on the others. And at the end of the calculations and observations it was noticed that Jupiter and Saturn went according

*Olaus Roemer, 1644–1710, Danish astronomer.

to the calculations, but that Uranus was doing something funny. Another opportunity for Newton's Laws to be found wanting; but take courage! Two men, Adams and Leverrier,* who made these calculations independently and at almost exactly the same time, proposed that the motions of Uranus were due to an unseen planet, and they wrote letters to their respective observatories telling them – 'Turn your telescope and look there and you will find a planet'. 'How absurd,' said one of the observatories, 'some guy sitting with pieces of paper and pencils can tell us where to look to find some new planet.' The other observatory was more . . . well, the administration was different, and they found Neptune!

More recently, in the beginning of the twentieth century, it became apparent that the motion of the planet Mercury was not exactly right. This caused a lot of trouble and was not explained until it was shown by Einstein that Newton's Laws were slightly off and that they had to be modified.

The question is, how far does this law extend? Does it extend outside the solar system? And so I show on Plate 1 evidence that the Law of Gravitation is on a wider scale than just the solar system. Here is a series of three pictures of a so-called double star. There is a third star fortunately in the picture so that you can see they are really turning around and that nobody simply turned the frames of the pictures around, which is easy to do on astronomical pictures. The stars are actually going around, and you can see the orbit that they make on figure 5. It is evident that they are attracting each other and that they are going around in an ellipse according to the way expected. These are a succession of positions at various times going around clockwise. You will be happy except when you notice, if you have not noticed already, that the centre is not a focus of the ellipse but is quite a bit off. So something is the matter with the law? No, God has not presented us with this orbit face-on; it is tilted

*John Couch Adams, 1819–92, mathematical astronomer. Urbain Leverrier, 1811–77, French astronomer.

21 July, 1908

September, 1915

10 July, 1920

Plate 1. Three photographs taken at different times of the same double star system.

Plate 2. A globular star cluster

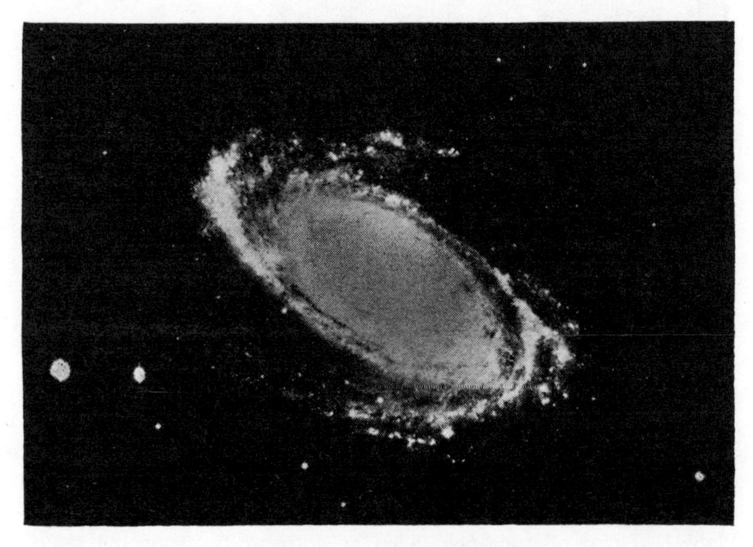

Plate 3. A spiral galaxy

Plate 4. A cluster of galaxies

Plate 5. A gaseous nebula

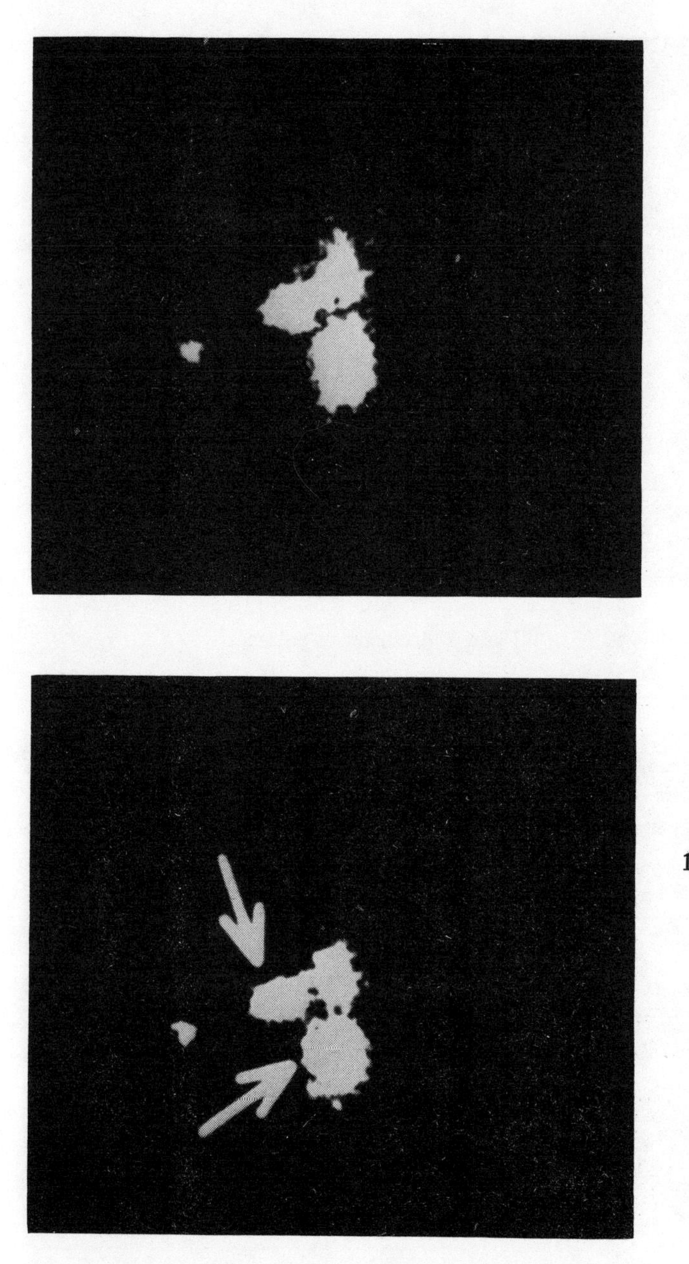

Plate 6. Evidence of the creation of new stars

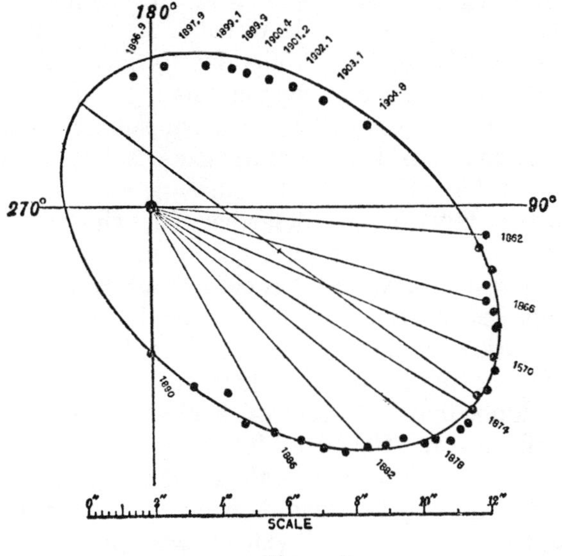

Figure 5

at a funny angle. If you take an ellipse and mark its focus and hold the paper at an odd angle and look at it in projection, you will find that the focus does not have to be at the focus of the projected image. It is because the orbit is tilted in space that it looks that way.

How about a bigger distance? This force is between two stars; does it go any farther than distances which are not more than two or three times the solar system's diameter? Here is something in plate 2 that is 100,000 times as big as the solar system in diameter; this is a tremendous number of stars. This large white spot is not a solid white spot; it appears like that because of the failure of the instruments to resolve it, but there are very very tiny spots just like other stars, well separated from each other, not hitting one another, each one falling through and back and forth in this great globular cluster. It is one of the most beautiful things in the sky; it is as beautiful as sea waves and sunsets. The distribution of this material is perfectly clear. The thing that holds this galaxy together is the gravitational attraction of the stars for each other. The distribution of the material and the sense of distance permits one to find out roughly what the law of force is between the stars . . . and, of course, it comes out that it is roughly the inverse square. Accuracy in these calculations and measurements is not anywhere near as careful as in the solar system.

Onward, gravity extends still farther. That cluster was just a little pin-point inside the big galaxy in plate 3, which shows a typical galaxy, and it is clear that again this thing is held together by some force, and the only candidate that is reasonable is gravitation. When we get to this size we have no way of checking the inverse square law, but there seems to be no doubt that in these great agglomerations of stars – these galaxies are 50,000 to 100,000 light years across, while the distance from the earth to the sun is only eight light minutes – gravity is extending even over these distances. In plate 4 is evidence that it extends even farther. This is what is called a cluster of galaxies; they are all in one lump and analogous to the cluster of stars, but this time what is clustered are those big babies shown in plate 3.

This is as far as about one tenth, maybe a hundredth, of the size of the Universe, as far as we have any direct evidence that gravitational forces extend. So the earth's gravitation has no edge, although you may read in the papers that something gets outside the field of gravitation. It becomes weaker and weaker inversely as the square of the distance, divided by four each time you get twice as far away, until it is lost in the confusion of the strong fields of other stars. Together with the stars in its neighbourhood it pulls the other stars to form the galaxy, and all together they pull on other galaxies and make a pattern, a cluster, of galaxies. So the earth's gravitational field never ends, but peters out very slowly in a precise and careful law, probably to the edges of the Universe.

The Law of Gravitation is different from many of the others. Clearly it is very important in the economy, in the machinery, of the Universe; there are many places where gravity has its practical applications as far as the Universe is concerned. But atypically the knowledge of the Laws of Gravitation has relatively few practical applications compared with the other laws of physics. This is one case where I have picked an atypical example. It is impossible, by the way, by picking one of anything to pick one that is not atypical in some sense. That is the wonder of the world. The only applications of the knowledge of the law that I can think of are in geophysical prospecting, in predicting the tides, and nowadays, more modernly, in working out the motions of the satellites and planet probes that we send up, and so on; and finally, also modernly, to calculate the predictions of the planets' positions, which have great utility for astrologists who publish their predictions in horoscopes in the magazines. It is a strange world we live in – that all the new advances in understanding are used only to continue the nonsense which has existed for 2,000 years.

I must mention the important places where gravitation does have some real effect in the behaviour of the Universe, and one of the interesting ones is in the formation of new stars. Plate 5 is a gaseous nebula inside our own galaxy; it is not a lot of stars; it is gas. The black specks are places where the gas has been compressed or attracted to itself. Perhaps it starts by some kind of shock waves, but the remainder of the phenomenon is that gravitation pulls the gas closer and closer together so that big mobs of gas and dust collect and form balls; and as they fall still farther, the heat generated by falling lights them up, and they become stars. And we have in plate 6 some evidence of the creation of new stars.

So this is how stars are born, when the gas collects together too much by gravitation. Sometimes when they explode the stars belch out dirt and gases, and the dirt and gases collect back again and make new stars – it sounds like perpetual motion.

I have already shown that gravitation extends to great distances, but Newton said that everything attracted everything else. Is it really true that two things attract each other?

Can we make a direct test and not just wait to see whether the planets attract each other? A direct test was made by Cavendish* on equipment which you see indicated in figure 6. The idea was to hang by a very very fine quartz fibre a

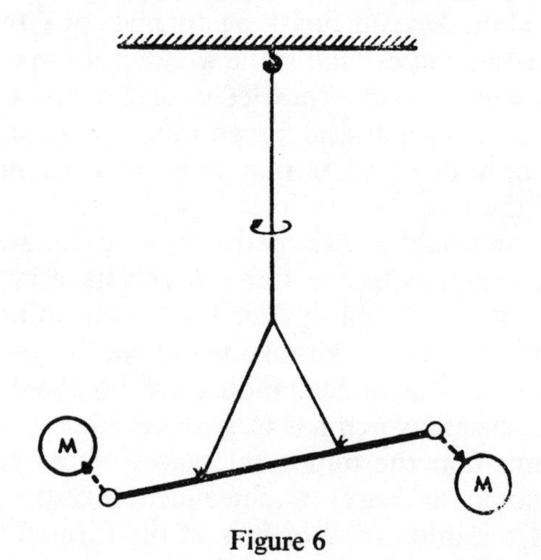

Figure 6

rod with two balls, and then put two large lead balls in the positions indicated next to it on the side. Because of the attraction of the balls there would be a slight twist to the fibre, and the gravitational force between ordinary things is very very tiny indeed. It was possible to measure the force between the two balls. Cavendish called his experiment 'weighing the earth'. With pedantic and careful teaching today we would not let our students say that; we would have to say 'measuring the mass of the earth'. By a direct experiment Cavendish was able to measure the force, the two masses and the distance, and thus determine the gravitational constant, G. You say, 'Yes, but we have the same situation here. We know what the pull is and we know what the mass of the object pulled is, and we know how far away we are, but we do not know either the mass of the earth or

*Henry Cavendish, 1731–1810, English physicist and chemist.

the constant, only the combination'. By measuring the constant, and knowing the facts about the pull of the earth, the mass of the earth could be determined.

Indirectly this experiment was the first determination of how heavy or massive is the ball on which we stand. It is an amazing achievement to find that out, and I think that is why Cavendish named his experiment 'weighing the earth', instead of 'determining the constant in the gravitational equation'. He, incidentally, was weighing the sun and everything else at the same time, because the pull of the sun is known in the same manner.

One other test of the law of gravity is very interesting, and that is the question whether the pull is exactly proportional to the mass. If the pull is exactly proportional to the mass, and the reaction to force, the motions induced by forces, changes in velocity, are inversely proportional to the mass. That means that two objects of different mass will change their velocity in the same manner in a gravitational field; or two different things in a vacuum, no matter what their mass, will fall the same way to the earth. That is Galileo's old experiment from the leaning tower of Pisa. It means, for example, that in a man-made satellite, an object inside will go round the earth in the same kind of orbit as one on the outside, and thus apparently float in the middle. The fact that the force is exactly proportional to the mass, and that the reactions are inversely proportional to the mass, has this very interesting consequence.

How accurate is it? It was measured in an experiment by a man named Eötvös* in 1909 and very much more recently and more accurately by Dicke,† and is known to one part in 10,000,000,000. The forces are exactly proportional to the mass. How is it possible to measure with that accuracy? Suppose you wanted to measure whether it is true for the pull of the sun. You know the sun is pulling us all, it pulls the earth too, but suppose you wanted to know whether the

*Baron Roland von Eötvös, 1848–1919, Hungarian physicist.
†Robert Henry Dicke, American physicist.

pull is exactly proportional to the inertia. The experiment was first done with sandalwood; lead and copper have been used, and now it is done with polyethylene. The earth is going around the sun, so the things are thrown out by inertia and they are thrown out to the extent that the two objects have inertia. But they are attracted to the sun to the extent that they have mass, in the attraction law. So if they are attracted to the sun in a different proportion from that thrown out by inertia, one will be pulled towards the sun, and the other away from it, and so, hanging them on opposite ends of a rod on another Cavendish quartz fibre, the thing will twist towards the sun. It does not twist at this accuracy, so we know that the sun's attraction to the two objects is exactly proportional to the centrifugal effect, which is inertia; therefore, the force of attraction on an object is exactly proportional to its coefficient of inertia; in other words, its mass.

One thing is particularly interesting. The inverse square law appears again – in the electrical laws, for instance. Electricity also exerts forces inversely as the square of the distance, this time between charges, and one thinks perhaps that the inverse square of the distance has some deep significance. No one has ever succeeded in making electricity and gravity different aspects of the same thing. Today our theories of physics, the laws of physics, are a multitude of different parts and pieces that do not fit together very well. We do not have one structure from which all is deduced; we have several pieces that do not quite fit exactly yet. That is the reason why in these lectures instead of having the ability to tell you what *the law* of physics is, I have to talk about the things that are common to the various laws; we do not understand the connection between them. But what is very strange is that there are certain things which are the same in both. Now let us look again at the law of electricity.

The force goes inversely as the square of the distance, but the thing that is remarkable is the tremendous difference in the strength of the electrical and gravitational forces. People who want to make electricity and gravitation out of the same thing will find that electricity is so much more powerful than gravity, it is hard to believe they could both have the same origin. How can I say one thing is more powerful than another? It depends upon how much charge you have, and how much mass you have. You cannot talk about how strong gravity is by saying: 'I take a lump of such a size', because *you* chose the size. If we try to get something that Nature produces – her own pure number that has nothing to do with inches or years or anything to do with our own dimensions – we can do it this way. If we take a fundamental particle such as an electron – any different one will give a different number, but to give an idea say electrons – two electrons are two fundamental particles, and they repel each other inversely as the square of the distance due to electricity, and they attract each other inversely as the square of the distance due to gravitation.

Question: What is the ratio of the gravitational force to the electrical force? That is illustrated in figure 7. The ratio

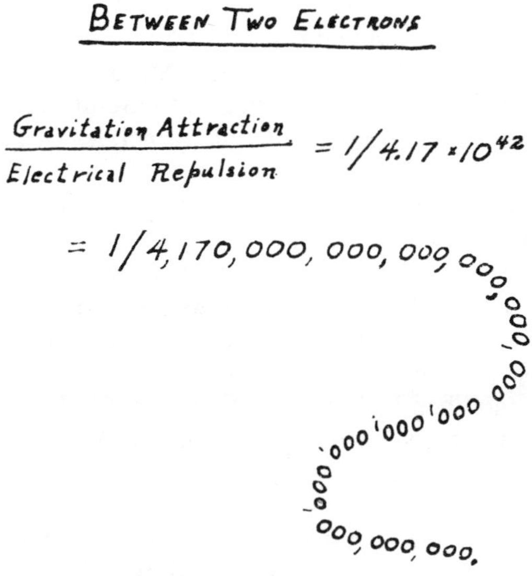

BETWEEN TWO ELECTRONS

$$\frac{\text{Gravitation Attraction}}{\text{Electrical Repulsion}} = 1/4.17 \times 10^{42}$$

$$= 1/4{,}170{,}000{,}000{,}000{,}000{,}000{,}000{,}000{,}000{,}000{,}000{,}000{,}000{,}000{,}000.$$

Figure 7

of the gravitational attraction to electrical repulsion is given by a number with 42 digits tailing off. Now therein lies a very deep mystery. Where could such a tremendous number come from? If you ever had a theory from which both of these things are to come, how could they come in such disproportion? What equation has a solution which has for two kinds of forces an attraction and repulsion with that fantastic ratio?

People have looked for such a large ratio in other places. They hope, for example, that there is another large number, and if you want a large number why not take the diameter of the Universe to the diameter of a proton – amazingly enough it also is a number with 42 digits. And so an interesting proposal is made that this ratio is the same as the ratio of the size of the Universe to the diameter of a proton. But the Universe is expanding with time and that means that the gravitational constant is changing with time, and although that is a possibility there is no evidence to indicate that it is a fact. There are several partial indications that the gravitational constant has not changed in that way. So this tremendous number remains a mystery.

To finish about the theory of gravitation, I must say two more things. One is that Einstein had to modify the Laws of Gravitation in accordance with his principles of relativity. The first of the principles was that 'x' cannot occur instantaneously, while Newton's theory said that the force was instantaneous. He had to modify Newton's laws. They have very small effects, these modifications. One of them is that all masses fall, light has energy and energy is equivalent to mass. So light falls and it means that light going near the sun is deflected; it is. Also the force of gravitation is slightly modified in Einstein's theory, so that the law has changed very very slightly, and it is just the right amount to account for the slight discrepancy that was found in the movement of Mercury.

Finally, in connection with the laws of physics on a small scale, we have found that the behaviour of matter on a small scale obeys laws very different from things on a large scale. So the question is, how does gravity look on a small scale? That is called the Quantum Theory of Gravity. There is no Quantum Theory of Gravity today. People have not succeeded completely in making a theory which is consistent with the uncertainty principles and the quantum mechanical principles.

You will say to me, 'Yes, you told us what happens, but what is gravity? Where does it come from? What is it? Do you mean to tell me that a planet looks at the sun, sees how far it is, calculates the inverse square of the distance and then decides to move in accordance with that law?' In other words, although I have stated the mathematical law, I have given no clue about the mechanism. I will discuss the possibility of doing this in the next lecture, 'The relation of mathematics to physics'.

In this lecture I would like to emphasize, just at the end, some characteristics that gravity has in common with the other laws that we mentioned as we passed along. First, it is mathematical in its expression; the others are that way too. Second, it is not exact; Einstein had to modify it, and we know it is not quite right yet, because we have still to put the quantum theory in. That is the same with all our other laws – they are not exact. There is always an edge of mystery, always a place where we have some fiddling around to do yet. This may or may not be a property of Nature, but it certainly is common to all the laws as we know them today. It may be only a lack of knowledge.

But the most impressive fact is that gravity is simple. It is simple to state the principles completely and not have left any vagueness for anybody to change the ideas of the law. It is simple, and therefore it is beautiful. It is simple in its pattern. I do not mean it is simple in its action – the motions of the various planets and the perturbations of one on the other can be quite complicated to work out, and to follow how all those stars in a globular cluster move is quite beyond our ability. It is complicated in its actions, but the basic pattern or the system beneath the whole thing is simple. This is common to all our laws; they all turn out to be simple things, although complex in their actual actions.

Finally comes the universality of the gravitational law, and the fact that it extends over such enormous distances that Newton, in his mind, worrying about the solar system, was able to predict what would happen in an experiment of Cavendish, where Cavendish's little model of the solar system, two balls attracting, has to be expanded ten million million times to become the solar system. Then ten million million times larger again we find galaxies attracting each other by exactly the same law. Nature uses only the longest threads to weave her patterns, so each small piece of her fabric reveals the organization of the entire tapestry.

2

The Relation of Mathematics to Physics

In thinking out the applications of mathematics and physics, it is perfectly natural that the mathematics will be useful when large numbers are involved in complex situations. In biology, for example, the action of a virus on a bacterium is unmathematical. If you watch it under a microscope, a jiggling little virus finds some spot on the odd shaped bacterium – they are all different shapes – and maybe it pushes its DNA in and maybe it does not. Yet if we do the experiment with millions and millions of bacteria and viruses, then we can learn a great deal about the viruses by taking averages. We can use mathematics in the averaging, to see whether the viruses develop in the bacteria, what new strains and what percentage; and so we can study the genetics, the mutations and so forth.

To take another more trivial example, imagine an enormous board, a chequerboard to play chequers or draughts. The actual operation of any one step is not mathematical – or it is very simple in its mathematics. But you could imagine that on an enormous board, with lots and lots of pieces, some analysis of the best moves, or the good moves or bad moves, might be made by a deep kind of reasoning which would involve somebody having gone off first and thought about it in great depth. That then becomes mathematics, involving abstract reasoning. Another example is switching in computers. If you have one switch, which is either on or off, there is nothing very mathematical about that, although mathematicians like to start there with their mathematics. But with all the interconnections and wires, to figure out what a very large system will do requires mathematics.

I would like to say immediately that mathematics has a tremendous application in physics in the discussion of the detailed phenomena in complicated situations, granting the fundamental rules of the game. That is something which I would spend most of my time discussing if I were talking only about the relation of mathematics and physics. But since this is part of a series of lectures on the character of physical law I do not have time to discuss what happens in complicated situations, but will go immediately to another question, which is the character of the fundamental laws.

If we go back to our chequer game, the fundamental laws are the rules by which the chequers move. Mathematics may be applied in the complex situation to figure out what in given circumstances is a good move to make. But very little mathematics is needed for the simple fundamental character of the basic laws. They can be simply stated in English for chequers.

The strange thing about physics is that for the fundamental laws we still need mathematics. I will give two examples, one in which we really do not, and one in which we do. First, there is a law in physics called Faraday's law, which says that in electrolysis the amount of material which is deposited is proportional to the current and to the time that the current is acting. That means that the amount of material deposited is proportional to the charge which goes through the system. It sounds very mathematical, but what is actually happening is that the electrons going through the wire each carry one charge. To take a particular example, maybe to deposit one atom requires one electron to come, so the number of atoms that are deposited is necessarily equal to the number of electrons that flow, and thus proportional to the charge that goes through the wire. So that mathematically-appearing law has as its basis nothing very deep, requiring no real knowledge of mathematics. That one electron is needed for each atom in order for it to deposit itself is mathematics, I suppose, but it is not the kind of mathematics that I am talking about here.

On the other hand, take Newton's law for gravitation,

which has the aspects I discussed last time. I gave you the equation:

$$F = G\frac{mm'}{r^2}$$

just to impress you with the speed with which mathematical symbols can convey information. I said that the force was proportional to the product of the masses of two objects, and inversely as the square of the distance between them, and also that bodies react to forces by changing their speeds, or changing their motions, in the direction of the force by amounts proportional to the force and inversely proportional to their masses. Those are words all right, and I did not necessarily have to write the equation. Nevertheless it is kind of mathematical, and we wonder how this can be a fundamental law. What does the planet do? Does it look at the sun, see how far away it is, and decide to calculate on its internal adding machine the inverse of the square of the distance, which tells it how much to move? This is certainly no explanation of the machinery of gravitation! You might want to look further, and various people have tried to look further. Newton was originally asked about his theory – 'But it doesn't mean anything – it doesn't tell us anything'. He said, 'It tells you *how* it moves. That should be enough. I have told you how it moves, not why.' But people often are unsatisfied without a mechanism, and I would like to describe one theory which has been invented, among others, of the type you might want. This theory suggests that this effect is the result of large numbers of actions, which would explain why it is mathematical.

Suppose that in the world everywhere there are a lot of particles, flying through us at very high speed. They come equally in all directions – just shooting by – and once in a while they hit us in a bombardment. We, and the sun, are practically transparent for them, practically but not completely, and some of them hit. Look, then, at what would happen (fig. 8).

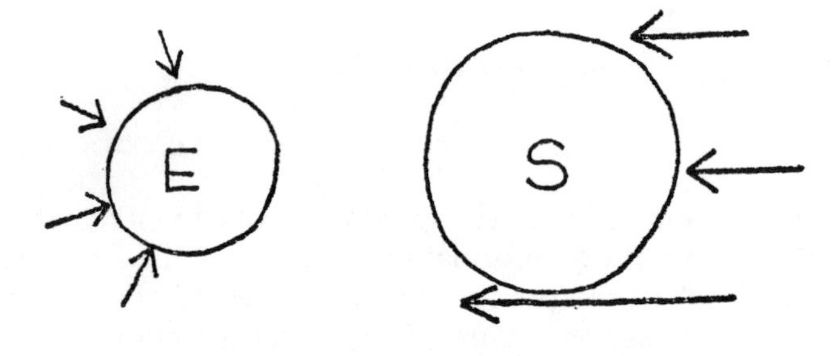

Figure 8

S is the sun, and E the earth. If the sun were not there, particles would be bombarding the earth from all sides, giving little impulses by the rattle, bang, bang of the few that hit. This will not shake the earth in any particular direction, because there are as many coming from one side as from the other, from top as from bottom. However, when the sun is there the particles which are coming from that direction are partly absorbed by the sun, because some of them hit the sun and do not go through. Therefore the number coming from the sun's direction towards the earth is less than the number coming from the other sides, because they meet an obstacle, the sun. It is easy to see that the farther the sun is away, of all the possible directions in which particles can come, a smaller proportion of the particles are being taken out. The sun will appear smaller – in fact inversely as the square of the distance. Therefore there will be an impulse on the earth towards the sun that varies inversely as the square of the distance. And this will be a result of large numbers of very simple operations, just hits, one after the other, from all directions. Therefore the strangeness of the mathematical relation will be very much reduced, because the fundamental operation is much simpler than calculating the inverse of the square of the distance. This design, with the particles bouncing, does the calculation.

The only trouble with this scheme is that it does not work, for other reasons. Every theory that you make up has to be analysed against *all* possible consequences, to see if it predicts anything else. And this does predict something else. If the earth is moving, more particles will hit it from in front than from behind. (If you are running in the rain, more rain hits you in the front of the face than in the back of the head, because you are running into the rain.) So, if the earth is moving it is running into the particles coming towards it and away from the ones that are chasing it from behind. So more particles will hit it from the front than from the back, and there will be a force opposing any motion. This force would slow the earth up in its orbit, and it certainly would not have lasted the three or four billion years (at least) that it has been going around the sun. So that is the end of that theory. 'Well,' you say, 'it was a good one, and I got rid of the mathematics for a while. Maybe I could invent a better one.' Maybe you can, because nobody knows the ultimate. But up to today, from the time of Newton, no one has invented another theoretical description of the mathematical machinery behind this law which does not either say the same thing over again, or make the mathematics harder, or predict some wrong phenomena. So there is no model of the theory of gravitation today, other than the mathematical form.

If this were the only law of this character it would be interesting and rather annoying. But what turns out to be true is that the more we investigate, the more laws we find, and the deeper we penetrate nature, the more this disease persists. Every one of our laws is a purely mathematical statement in rather complex and abstruse mathematics. Newton's statement of the law of gravitation is relatively simple mathematics. It gets more and more abstruse and more and more difficult as we go on. Why? I have not the slightest idea. It is only my purpose here to tell you about this fact. The burden of the lecture is just to emphasize he fact that it is impossible to explain honestly the beauties of the laws of nature in a way that people can feel, without their having some deep understanding of mathematics. I am sorry, but this seems to be the case.

You might say, 'All right, then if there is no explanation the of law, at least tell me what the law *is*. Why not tell me in words instead of in symbols? Mathematics is just a language, and I want to be able to translate the language'. In fact I can, with patience, and I think I partly did. I could go a little further and explain in more detail that the equation means that if the distance is twice as far the force is one fourth as much, and so on. I could convert all the symbols into words. In other words I could be kind to the laymen as they all sit hopefully waiting for me to explain something. Different people get different reputations for their skill at explaining to the layman in layman's language these difficult and abstruse subjects. The layman then searches for book after book in the hope that he will avoid the complexities which ultimately set in, even with the best expositor of this type. He finds as he reads a generally increasing confusion, one complicated statement after another, one difficult-to-understand thing after another, all apparently disconnected from one another. It becomes obscure, and he hopes that maybe in some other book there is some explanation. . . . The author almost made it – maybe another fellow will make it right.

But I do not think it is possible, because mathematics is *not* just another language. Mathematics is a language plus reasoning; it is like a language plus logic. Mathematics is a tool for reasoning. It is in fact a big collection of the results of some person's careful thought and reasoning. By mathematics it is possible to connect one statement to another. For instance, I can say that the force is directed towards the sun. I can also tell you, as I did, that the planet moves so that if I draw a line from the sun to the planet, and draw another line at some definite period, like three weeks, later, then the area that is swung out by the planet is exactly the same as it will be in the next three weeks, and the next three weeks, and so on as it goes around the sun. I can explain both of those statements carefully, but I cannot explain why they are both the same. The apparent enormous complexities of nature, with all its funny laws and rules, each of which has been carefully explained to you, are really very closely interwoven. However, if you do not appreciate the mathematics, you cannot see, among the great variety of facts, that logic permits you to go from one to the other.

It may be unbelievable that I can demonstrate that equal areas will be swept out in equal times if the forces are directed towards the sun. So if I may, I will do one demonstration to show you that those two things really are equivalent, so that you can appreciate more than the mere statement of the two laws. I will show that the two laws are connected so that reasoning alone will bring you from one to the other, and that mathematics is just organized reasoning. Then you will appreciate the beauty of the relationship of the statements. I am going to prove the relationship that if the forces are directed towards the sun equal areas are swept out in equal times.

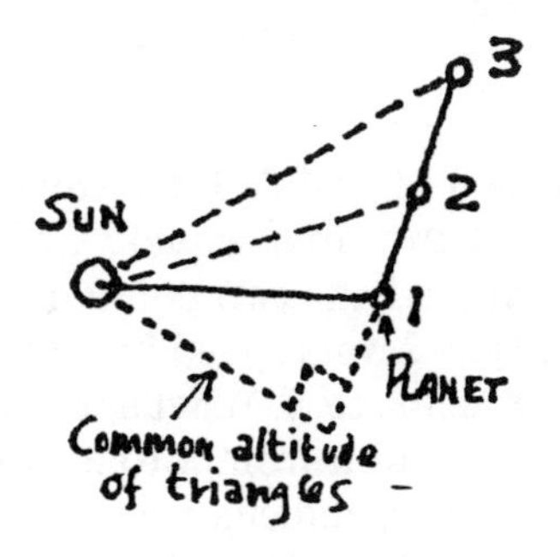

Figure 9

We start with a sun and a planet (fig. 9), and we imagine that at a certain time the planet is at position 1. It is moving in such a way that, say, one second later it has moved to position 2. If the sun did not exert a force on the planet, then, by Galileo's principle of inertia, it would keep right on going in a straight line. So after the same interval of time, the next second, it would have moved exactly the same distance in the same straight line, to the position 3. First we are going to show that if there is *no* force, then equal areas are swept out in equal times. I remind you that the area of a triangle is half the base times the altitude, and that the altitude is the vertical distance to the base. If the triangle is obtuse (fig. 10), then the altitude is the vertical height AD and the base is BC. Now let us compare the areas which would be swept out if the sun exerted no force whatsoever (fig. 9).

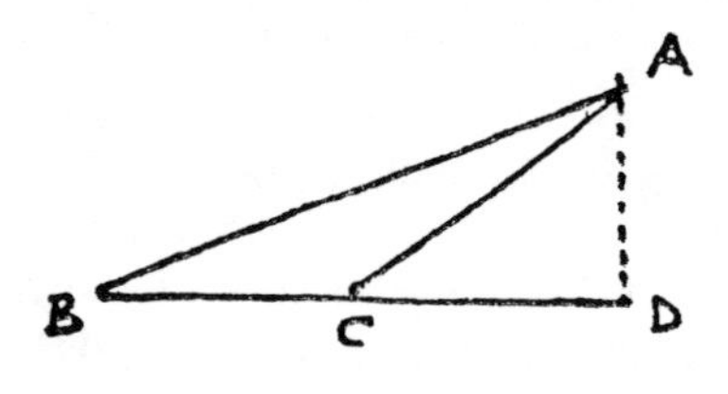

Figure 10

The two distances 1–2 and 2–3 are equal, remember. The question is, are the two areas equal? Consider the triangle made from the sun and the two points 1 and 2. What is its area? It is the base 1–2, multiplied by half the perpendicular height from the baseline to S. What about the other triangle, the triangle in the motion from 2 to 3? Its area is the base 2–3, times half the perpendicular height to S. The two triangles have the same altitude, and, as I indicated, the same base, and therefore they have the same area. So far so good. If there were no force from the sun, equal areas would be swept out in equal times. But there *is* a force from the sun. During the interval 1–2–3 the sun is pulling and changing the motion in various directions towards itself. To get a good approximation we will take the central position, or average position, at 2, and say that the whole effect during the interval 1–3 was to change the motion by some amount in the direction of the line 2–S (fig. 11).

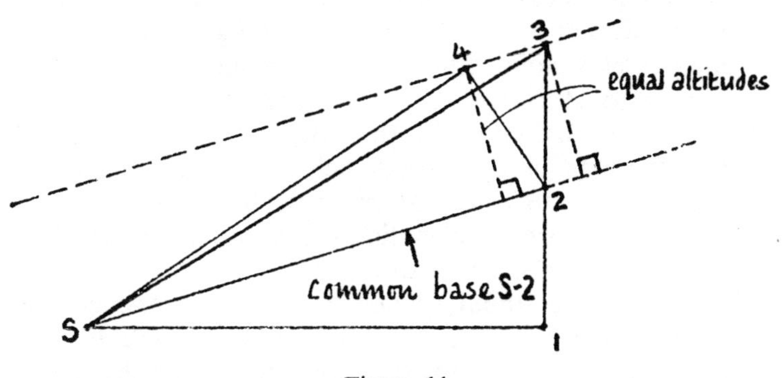

Figure 11

This means that though the particles were moving on the line 1–2, and would, were there no force, have continued to move on the same line in the next second, because of the influence of the sun the motion is altered by an amount that is poking in a direction parallel to the line 2–S. The next motion is therefore a compound of what the planet wanted to do and the change that has been induced by the action of the sun. So the planet does not really end up at position 3, but rather at position 4. Now we would like to compare the areas of the triangles 23S and 24S, and I will show you that those are equal. They have the same base, S–2. Do they have the same altitude? Sure, because they are included between parallel lines. The distance from 4 to the line S–2 is equal to the distance from 3 to line S–2 (extended). Thus the area of the triangle S24 is the same as S23. I proved earlier that S12 and S23 were equal in area, so we now know S12 = S24. So, in the actual orbital motion of the planet the areas swept out in the first second and the second second are equal. Therefore, by reasoning, we can see a connection between the fact that the force is towards the sun, and the fact that the areas are equal. Isn't that ingenious? I borrowed it straight from Newton. It comes right out of the *Principia*, diagram and all. Only the letters are different, because he wrote in Latin and these are Arabic numerals.

Newton made all the proofs in his book geometrical. Today we do not use that kind of reasoning. We use a kind of analytic reasoning with symbols. It requires ingenuity to draw the correct triangles, to notice about the areas, and to figure out how to do this. But there have been improvements in the methods of analysis, which are faster and more efficient. I want to show what this looks like in the notation of the more modern mathematics, where you do nothing but write a lot of symbols to figure it out.

We want to talk about how fast the area changes, and we represent that by $\dot{A}$. The area changes when the radius is swinging, and it is the component of velocity at right angles to the radius, times the radius, that tells us how fast the area changes. So this is the component of the radial distance multiplied by the velocity, or rate of change of the distance.

$$\dot{A} = \vec{r} \times \dot{\vec{r}}$$

The question now is whether the rate of change of area itself changes. The principle is that the rate of change of the area is not supposed to change. So we differentiate this again, and this means some little trick about putting dots in the right place, that is all. You have to learn the tricks; it is just a series of rules that people have found out that are very useful for such a thing. We write:

$$\ddot{A} = \dot{\vec{r}} \times \dot{\vec{r}} + \vec{r} \times \ddot{\vec{r}} = \vec{r} \times \vec{F}/m$$

This first term says to take the component of the velocity at right angles to the velocity. It is zero; the velocity is in the same direction as itself. The acceleration, which is the second derivative, r with two dots, or the derivative of the velocity, is the force divided by the mass.

This says therefore that the rate of change of the rate of change of the area is the component of force at right angles to the radius, but if the force is in the direction of the radius,

$$\vec{r} \times \vec{F}/m = 0 \text{ or } \ddot{A} = 0$$

as Newton said, then there is no force at right angles to the radius, and that means that the rate of change of area does not change. This merely illustrates the power of analysis with different kinds of notation. Newton knew how to do this, more or less, with slightly different notations; but he wrote everything in the geometrical form, because he tried to make it possible for people to read his papers. He invented the calculus, which is the kind of mathematics I have just shown.

This is a good illustration of the relation of mathematics to physics. When the problems in physics become difficult we may often look to the mathematicians, who may already have studied such things and have prepared a line of reasoning for us to follow. On the other hand they may not have, in which case we have to invent our own line of reasoning, which we then pass back to the mathematicians. Everybody who reasons carefully about anything is making a contribution to the knowledge of what happens when you think about something, and if you abstract it away and send it to the Department of Mathematics they put it in books as a branch of mathematics. Mathematics, then, is a way of going from one set of statements to another. It is evidently useful in physics, because we have these different ways in which we can speak of things, and mathematics permits us to develop consequences, to analyse the situations, and to change the laws in different ways to connect the various statements. In fact the total amount that a physicist knows is very little. He has only to remember the rules to get him from one place to another and he is all right, because all the various statements about equal times, the force being in the direction of the radius, and so on, are all interconnected by reasoning.

Now an interesting question comes up. Is there a place to begin to deduce the whole works? Is there some particular pattern or order in nature by which we can understand that one set of statements is more fundamental and one set of statements more consequential? There are two kinds of ways of looking at mathematics, which for the purpose of this lecture I will call the Babylonian tradition and the Greek tradition. In Babylonian schools in mathematics the student would learn something by doing a large number of examples until he caught on to the general rule. Also he would know a large amount of geometry, a lot of the properties of circles, the theorem of Pythagoras, formulae for the areas of cubes and triangles; in addition, some degree of argument was available to go from one thing to another. Tables of numerical quantities were available so that they could solve elaborate equations. Everything was prepared for calculating things out. But Euclid discovered that there was a way in which all of the theorems of geometry could be ordered from a set of axioms that were particularly simple. The Babylonian attitude – or what I call Babylonian mathematics – is that you know all of the various theorems and many of the connections in between, but you have never fully realized that it could all come up from a bunch of axioms. The most modern mathematics concentrates on axioms and demonstrations within a very definite framework of conventions of what is acceptable and what is not acceptable as axioms. Modern geometry takes something like Euclid's axioms, modified to be more perfect, and then shows the deduction of the system. For instance, it would not be expected that a theorem like Pythagoras's (that the sum of the areas of squares put on two sides of a right-angled triangle is equal to the area of the square on the hypotenuse) should be an axiom. On the other hand, from another point of view of geometry, that of Descartes, the Pythagorean theorem is an axiom.

So the first thing we have to accept is that even in mathematics you can start in different places. If all these various theorems are interconnected by reasoning there is no real way to say 'These are the most fundamental axioms', because if you were told something different instead you could also run the reasoning the other way. It is like a bridge with lots of members, and it is over-connected; if pieces have dropped out you can reconnect it another way. The mathematical tradition of today is to start with some particular ideas which are chosen by some kind of convention to be axioms, and then to build up the structure from there. What I have called the Babylonian idea is to say, 'I happen to know this, and I happen to know that, and maybe I know that; and I work everything out from there. Tomorrow I may forget that this is true, but remember that something else is true, so I can reconstruct it all again. I am never quite sure of where I am supposed to begin or where I am supposed to end. I just remember enough all the time so that as the memory fades and some of the pieces fall out I can put the thing back together again every day'.

The method of always starting from the axioms is not very efficient in obtaining theorems. In working something out in geometry you are not very efficient if each time you have to start back at the axioms. If you have to remember a few things in geometry you can always get somewhere else, but it is much more efficient to do it the other way. To decide which are the best axioms is not necessarily the most efficient way of getting around in the territory. In physics we need the Babylonian method, and not the Euclidian or Greek method. I would like to explain why.

The problem in the Euclidian method is to make something about the axioms a little more interesting or important. But in the case of gravitation, for example, the question we are asking is: is it more important, more basic, or is it a better axiom, to say that the force is towards the sun, or to say that equal areas are swept out in equal times? From one point of view the force statement is better. If I state what the forces are I can deal with a system with many particles in which the orbits are no longer ellipses, because the force statement tells me about the pull of one on the other. In this case the theorem about equal areas fails. Therefore I think that the force law ought to be an axiom instead of the other. On the other hand, the principle of equal areas can be generalized, in a system of a large number of particles, to another theorem. It is rather complicated to say, and not quite as pretty as the original statement about equal areas, but it is obviously its offspring. Take a system with a large number of particles, perhaps Jupiter, Saturn, the Sun, and lots of stars, all interacting with each other, and look at it from far away projected on a plane (fig. 12). The particles are all moving in various directions, and we take any point and calculate how much area is being swept out by the radius from this point to each of the particles. In this calculation the masses which are heavier count more strongly; if one particle is twice as heavy as another its area will count twice as much. So we count each of the areas swept out in proportion to the mass that is doing the sweeping, add them all together, and the *resulting total is not changing in time.* That total is called the angular momentum, and this is called the law of conservation of angular momentum. Conservation just means that it does not change.

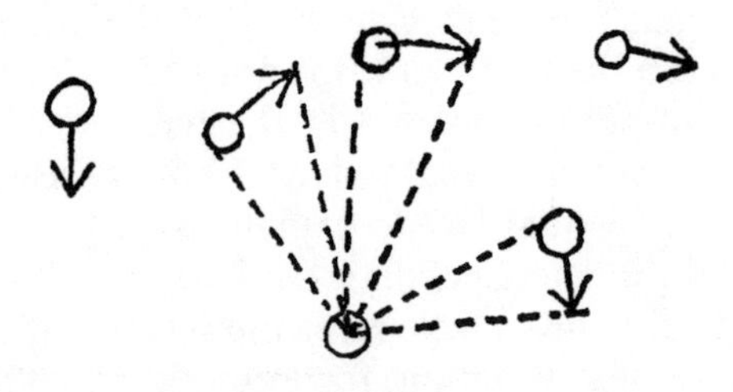

Figure 12

One of the consequences of this is as follows. Imagine a lot of stars falling together to form a nebula, or galaxy. At first they are very far out, on long radii from the centre, moving slowly and allowing a small amount of area to be generated. As they come closer the distances to the centre will shorten, and when they are very far in the radii will be very small, so in order to produce the same area per second they will have to move a great deal faster. You will see then that as the stars come in they will swing and swirl around faster and faster, and thus we can roughly understand the qualitative shape of the spiral nebulae. In the same way we can understand how a skater spins. He starts with his leg out, moving slowly, and as he pulls his leg in he spins faster. When the leg is out it is contributing a certain amount of area per second, and then when he brings his leg in he has to spin much faster to produce the same amount of area. But I did not prove it for the skater: the skater uses muscle force, and gravity is a different force. Yet it *is* true for the skater.

Now we have a problem. We can deduce often from one part of physics, like the Law of Gravitation, a principle which turns out to be much more valid than the derivation. This does not happen in mathematics; theorems do not come out in places where they are not supposed to be. In other words, if we were to say that the postulate of physics was the equal area law of gravitation, then we could deduce the conservation of angular momentum, but only for gravitation. Yet we discover experimentally that the conservation of angular momentum is a much wider thing. Newton had other postulates by which he could get the more general conservation law of angular momentum. But these Newtonian laws were wrong. There are no forces, it is all a lot of boloney, the particles do not have orbits, and so on. Yet the analogue, the exact transformation of this principle about the areas and the conservation of angular momentum, is true. It works for atomic motions in quantum mechanics, and, as far as we can tell, it is still exact today. We have these wide principles which sweep across the different laws, and if we take the derivation too seriously, and feel that one is only valid because another is valid, then we cannot understand the interconnections of the different branches of physics. Some day, when physics is complete and we know all the laws, we may be able to start with some axioms, and no doubt somebody will figure out a particular way of doing it so that everything else can be deduced. But while we do not know all the laws, we can use some to make guesses at theorems which extend beyond the proof. In order to understand physics one must always have a neat balance, and contain in one's head all of the various propositions and their interrelationships, because the laws often extend beyond the range of their deductions. This will only have no importance when all the laws are known.

Another thing, a very strange one, that is interesting in the relation of mathematics to physics is the fact that by mathematical arguments you can show that it is possible to start from many apparently different starting points, and yet come to the same thing. That is pretty clear. If you have axioms, you can instead use some of the theorems; but actually the physical laws are so delicately constructed that the different but equivalent statements of them have such qualitatively different characters, and this makes them very interesting. To illustrate this I am going to state the law of gravitation in three different ways, all of which are exactly equivalent but sound completely different.

The first statement is that there are forces between objects, according to the equation which I have given you before.

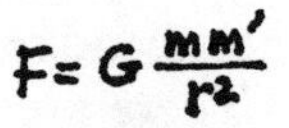

Each object, when it sees the force on it, accelerates or changes its motion, at a certain amount per second. It is the regular way of stating the law, I call it Newton's law. This statement of the law says that the force depends on something at a finite distance away. It has what we call an unlocal quality. The force on one object depends on where another one is some distance away.

You may not like the idea of action at a distance. How can this object know what is going on over there? So there is another way of stating the laws, which is very strange, called the field way. It is hard to explain, but I want to give you some rough idea of what it is like. It says a completely different thing. There is a number at every point in space (I know it is a number, not a mechanism: that is the trouble with physics, it must be mathematical), and the numbers change when you go from place to place. If an object is placed at a point in space, the force on it is in the direction in which that number changes most rapidly (I will give it its usual name, the potential, the force is in the direction in which the potential changes). Further, the force is proportional to how fast the potential changes as you move. That is one part of the statement, but it is not enough, because I have yet to tell you how to determine the way in which the potential varies. I could say the potential varies inversely as the distance from each object, but that is back to the reaction-at-a-distance idea. You can state the law in another way, which says that you do not have to know what is going on anywhere outside a little ball. If you want to know what the potential is at the centre of the ball, you need only tell me what it is on the surface of the ball, however small. You do not have to look outside, you just tell me what it is in the neighbourhood, and how much mass there is in the ball. The rule is this. The potential at the centre is equal to the average of the potential on the surface of the ball, minus the same constant, G, as we had in the other equation, divided by twice the radius of the ball (which we will call a), and then multiplied by the mass inside the ball, if the ball is small enough.

$$\text{Potential at centre} = \text{Av. pot. on ball} - \frac{G}{2a}(\text{mass inside})$$

You see that this law is different from the other, because it tells what happens at one point in terms of what happens very close by. Newton's law tells what happens at one time in terms of what happens at another instant. It gives from instant to instant how to work it out, but in space leaps from place to place. The second statement is both local in time and local in space, because it depends only on what is in the neighbourhood. But both statements are exactly equivalent mathematically.

There is another completely different way of stating this, different in the philosophy and the qualitative ideas involved. If you do not like action at a distance I have shown you can get away without it. Now I want to show you a statement which is philosophically the exact opposite. In this there is no discussion at all about how the thing works its way from place to place; the whole is contained in an overall statement, as follows. When you have a number of particles, and you want to know how one moves from one place to another, you do it by inventing a possible motion that gets from one place to the other in a given amount of time (fig. 13). Say the particle wants to go from X to Y in an hour, and you want to know by what route it can go. What you do is to invent various curves, and calculate on each curve a certain quantity. (I do not want to tell you what the quantity is, but for those who have heard of these terms the quantity on each route is the average of the difference between the kinetic and the potential energy.) If you calculate this quantity for one route, and then for another, you will get a different number for each route. There is one route which gives the least possible number, however, and that is the route that the particle in nature actually takes! We are now describing the actual motion, the ellipse, by saying something about the whole curve. We have lost the idea of causality, that the particle feels the pull and moves in accordance with it. Instead of that, in some grand fashion it smells all the curves, all the possibilities, and decides which one to take (by choosing that for which our quantity is least).

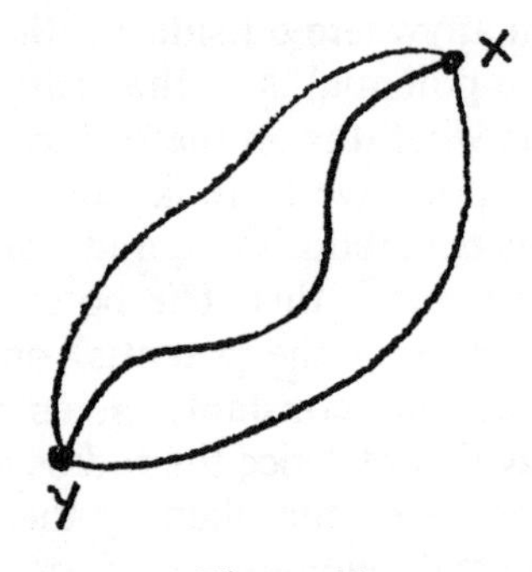

Figure 13

This is an example of the wide range of beautiful ways of describing nature. When people say that nature must have causality, you can use Newton's law; or if they say that nature must be stated in terms of a minimum principle, you talk about it this last way; or if they insist that nature must have a local field – sure, you can do that. The question is: which one is right? If these various alternatives are not exactly equivalent mathematically, if for certain ones there will be different consequences than for others, then all we have to do is to experiment to find out which way nature actually chooses to do it. People may come along and argue philosophically that they like one better than another; but we have learned from much experience that all philosophical intuitions about what nature is going to do fail. One just has to work out all the possibilities, and try all the alternatives. But in the particular case I am talking about the theories *are* exactly equivalent. Mathematically each of the three different formulations, Newton's law, the local field method and the minimum principle, gives exactly the same consequences. What do we do then? You will read in all the books that we cannot decide scientifically on one or the other. That is true. They are equivalent scientifically. It is impossible to make a decision, because there is no experimental way to distinguish between them if all the consequences are the same. But psychologically they are very different in two ways. First, philosophically you like them or do not like them; and training is the only way to beat that disease. Second, psychologically they are different because they are completely unequivalent when you are trying to guess new laws.

As long as physics is incomplete, and we are trying to understand the other laws, then the different possible formulations may give clues about what might happen in other circumstances. In that case they are no longer equivalent, psychologically, in suggesting to us guesses about what the laws may look like in a wider situation. To give an example, Einstein realized that *electrical* signals could not propagate faster than the speed of light. He guessed that it was a general principle. (This is the same guessing game as taking the angular momentum and extending it from one case where you have proved it, to the rest of the phenomena of the universe.) He guessed that it was true of everything, and he guessed that it would be true of gravitation. If signals cannot go any faster than the speed of light, then it turns out that the method of describing the forces instantaneously is very poor. So in Einstein's generalization of gravitation Newton's method of describing physics is hopelessly inadequate and enormously complicated, whereas the field method is neat and simple, and so is the minimum principle. We have not decided between the last two yet.

In fact it turns out that in quantum mechanics neither is right in exactly the way I have stated them, but the fact that a minimum principle exists turns out to be a consequence of the fact that on a small scale particles obey quantum mechanics. The best law, as at present understood, is really a combination of the two in which we use minimum principles plus local laws. At present we believe that the laws of physics have to have the local character and also the minimum principle, but we do not really know. If you have a structure that is only partly accurate, and something is going to fail, then if you write it with just the right axioms maybe only one axiom fails and the rest remain, you need only change one little thing. But if you write it with another set of axioms they may all collapse, because they all lean on that one thing that fails. We cannot tell ahead of time, without some intuition, which is the best way to write it so that we can find out the new situation. We must always keep all the alternative ways of looking at a thing in our heads; so physicists do Babylonian mathematics, and pay but little attention to the precise reasoning from fixed axioms.

One of the amazing characteristics of nature is the variety of interpretational schemes which is possible. It turns out that it is only possible because the laws are just so, special and delicate. For instance, that the law is the inverse square is what permits it to become local; if it were the inverse cube it could not be done that way. At the other end of the equation, the fact that the force is related to the rate of change of velocity is what permits the minimum principle way of writing the laws. If, for instance, the force were proportional to the rate of change of position instead of velocity, then you could not write it in that way. If you modify the laws much you find that you can only write them in fewer ways. I always find that mysterious, and I do not understand the reason why it is that the correct laws of physics seem to be expressible in such a tremendous variety of ways. They seem to be able to get through several wickets at the same time.

I should like to say a few things on the relation of mathematics and physics which are a little more general. Mathematicians are only dealing with the structure of reasoning, and they do not really care what they are talking about. They do not even need to *know* what they are talking about, or, as they themselves say, whether what they say is true. I will explain that. You state the axioms, such-and-such is so, and such-and-such is so. What then? The logic can be carried out without knowing what the such-and-such words mean. If the statements about the axioms are carefully formulated and complete enough, it is not necessary for the man who is doing the reasoning to have any knowledge of the meaning of the words in order to deduce new conclusions in the same language. If I use the word triangle in one of the axioms there will be a statement about triangles in the conclusion, whereas the man who is doing the reasoning may not know what a triangle is. But I can read his reasoning back and say, 'Triangle, that is just a three-sided what-have-you, which is so-and-so', and then I know his new facts. In other words, mathematicians prepare abstract reasoning ready to be used if you have a set of axioms about the real world. But the physicist has meaning to all his phrases. That is a very important thing that a lot of people who come to physics by way of mathematics do not appreciate. Physics is not mathematics, and mathematics is not physics. One helps the other. But in physics you have to have an understanding of the connection of words with the real world. It is necessary at the end to translate what you have figured out into English, into the world, into the blocks of copper and glass that you are going to do the experiments with. Only in that way can you find out whether the consequences are true. This is a problem which is not a problem of mathematics at all.

Of course it is obvious that the mathematical reasonings which have been developed are of great power and use for physicists. On the other hand, sometimes the physicists' reasoning is useful for mathematicians.

Mathematicians like to make their reasoning as general as possible. If I say to them, 'I want to talk about ordinary three dimensional space', they say 'If you have a space of n dimensions, then here are the theorems'. 'But I only want the case 3', 'Well, substitute $n = 3$.'! So it turns out that many of the complicated theorems they have are much simpler when adapted to a special case. The physicist is always interested in the special case; he is never interested in the general case. He is talking about something; he is not talking abstractly about anything. He wants to discuss the gravity law in three dimensions; he never wants the arbitrary force case in n dimensions. So a certain amount of reducing is necessary, because the mathematicians have prepared these things for a wide range of problems. This is very useful, and later on it always turns out that the poor physicist has to come back and say, 'Excuse me, when you wanted to tell me about four dimensions . . .'

When you know what it is you are talking about, that some symbols represent forces, others masses, inertia, and so on, then you can use a lot of commonsense, seat-of-the-pants feeling about the world. You have seen various things, and you know more or less how the phenomenon is going to behave. But the poor mathematician translates it into equations, and as the symbols do not mean anything to him he has no guide but precise mathematical rigour and care in the argument. The physicist, who knows more or less how the answer is going to come out, can sort of guess part way, and so go along rather rapidly. The mathematical rigour of great precision is not very useful in physics. But one should not criticize the mathematicians on this score. It is not necessary that just because something would be useful to physics they have to do it that way. They are doing their own job. If you want something else, then you work it out for yourself.

The next question is whether, when trying to guess a new law, we should use the seat-of-the-pants feeling and philosophical principles – 'I don't like the minimum principle', or 'I do like the minimum principle', 'I don't like action at a distance', or 'I do like action at a distance'. To what extent do models help? It is interesting that very often models do help, and most physics teachers try to teach how to use models and to get a good physical feel for how things are going to work out. But it always turns out that the greatest discoveries abstract away from the model and the model never does any good. Maxwell's discovery of electrodynamics was first made with a lot of imaginary wheels and idlers in space. But when you get rid of all the idlers and things in space the thing is O.K. Dirac* discovered the correct laws for relativity quantum mechanics simply by guessing the equation. The method of guessing the equation seems to be a pretty effective way of guessing new laws. This shows again that mathematics is a deep way of expressing nature, and any attempt to express nature in philosophical principles, or in seat-of-the-pants mechanical feelings, is not an efficient way.

It always bothers me that, according to the laws as we understand them today, it takes a computing machine an infinite number of logical operations to figure out what goes on in no matter how tiny a region of space, and no matter how tiny a region of time. How can all that be going on in that tiny space? Why should it take an infinite amount of logic to figure out what one tiny piece of space/time is going to do? So I have often made the hypothesis that

*Paul Dirac, British physicist. Joint Nobel Prize with Schrödinger, 1933.

ultimately physics will not require a mathematical statement, that in the end the machinery will be revealed, and the laws will turn out to be simple, like the chequer board with all its apparent complexities. But this speculation is of the same nature as those other people make – 'I like it', 'I don't like it', – and it is not good to be too prejudiced about these things.

To summarize, I would use the words of Jeans, who said that 'the Great Architect seems to be a mathematician'. To those who do not know mathematics it is difficult to get across a real feeling as to the beauty, the deepest beauty, of nature. C. P. Snow talked about two cultures. I really think that those two cultures separate people who have and people who have not had this experience of understanding mathematics well enough to appreciate nature once.

It is too bad that it has to be mathematics, and that mathematics is hard for some people. It is reputed – I do not know if it is true – that when one of the kings was trying to learn geometry from Euclid he complained that it was difficult. And Euclid said, 'There is no royal road to geometry'. And there *is* no royal road. Physicists cannot make a conversion to any other language. If you want to learn about nature, to appreciate nature, it is necessary to understand the language that she speaks in. She offers her information only in one form; we are not so unhumble as to demand that she change before we pay any attention.

All the intellectual arguments that you can make will not communicate to deaf ears what the experience of music really is. In the same way all the intellectual arguments in the world will not convey an understanding of nature to those of 'the other culture'. Philosophers may try to teach you by telling you qualitatively about nature. I am trying to describe her. But it is not getting across because it is impossible. Perhaps it is because their horizons are limited in this way that some people are able to imagine that the centre of the universe is man.

3

The Great Conservation Principles

When learning about the laws of physics you find that there are a large number of complicated and detailed laws, laws of gravitation, of electricity and magnetism, nuclear interactions, and so on, but across the variety of these detailed laws there sweep great general principles which all the laws seem to follow. Examples of these are the principles of conservation, certain qualities of symmetry, the general form of quantum mechanical principles, and unhappily, or happily, as we considered last time, the fact that all the laws are mathematical. In this lecture I want to talk about the conservation principles.

The physicist uses ordinary words in a peculiar manner. To him a conservation law means that there is a number which you can calculate at one moment, then as nature undergoes its multitude of changes, if you calculate this quantity again at a later time it will be the same as it was before, the number does not change. An example is the conservation of energy. There is a quantity that you can calculate according to a certain rule, and it comes out the same answer always, no matter what happens.

Now you can see that such a thing is possibly useful. Suppose that physics, or rather nature, is considered analogous to a great chess game with millions of pieces in it, and we are trying to discover the laws by which the pieces move. The great gods who play this chess play it very rapidly, and it is hard to watch and difficult to see. However, we are catching on to some of the rules, and there are some rules which we can work out which do not require that we watch every move. For instance, suppose there is one bishop only, a red bishop, on the board, then since the bishop moves diagonally and therefore never changes the colour of its square, if we look away for a moment while the gods play and then look back again, we can expect that there will be still a red bishop on the board, maybe in a different place, but on the same colour square. This is in the nature of a conservation law. We do not need to watch the insides to know at least something about the game.

It is true that in chess this particular law is not necessarily perfectly valid. If we looked away long enough it could happen that the bishop was captured, a pawn went down to queen, and the god decided that it was better to hold a bishop instead of a queen in the place of that pawn, which happened to be on a black square. Unfortunately it may well turn out that some of the laws which we see today may not be exactly perfect, but I will tell you about them as we see them at present.

I have said that we use ordinary words in a technical fashion, and another word in the title of this lecture is 'great', 'The Great Conservation Principles'. This is not a technical word: it was merely put in to make the title sound more dramatic, and I could just as well have called it 'The Conservation Laws'. There are a few conservation laws that do not work; they are only approximately right, but are sometimes useful, and we might call those the 'little' conservation laws. I will mention later one or two of those that do not work, but the principal ones that I am going to discuss are, as far as we can tell today, absolutely accurate.

I will start with the easiest one to understand, and that is the conservation of electric charge. There is a number, the total electric charge in the world, which, no matter what happens, does not change. If you lose it in one place you will find it in another. The conservation is of the total of all electric charge. This was discovered experimentally by Faraday.* The experiment consisted of getting inside a great globe of metal, on the outside of which was a very delicate galvanometer, to look for the charge on the globe,

*Michael Faraday, 1791–1867, English physicist.

because a small amount of charge would make a big effect. Inside the globe Faraday built all kinds of weird electrical equipment. He made charges by rubbing glass rods with cat's fur, and he made big electrostatic machines so that the inside of this globe looked like those horror movie laboratories. But during all these experiments no charge developed on the surface; there was no net charge made. Although the glass rod may have been positive after it was charged up by rubbing on the cat's fur, then the fur would be the same amount negative, and the total charge was always nothing, because if there were any charge developed on the inside of the globe it would have appeared as an effect in the galvanometer on the outside. So the total charge is conserved.

This is easy to understand, because a very simple model, which is not mathematical at all, will explain it. Suppose the world is made of only two kinds of particles, electrons and protons – there was a time when it looked as if it was going to be as easy as that – and suppose that the electrons carry a negative charge and the protons a positive charge, so that we can separate them. We can take a piece of matter and put on more electrons, or take some off; but supposing that electrons are permanent and never disintegrate or disappear – that is a simple proposition, not even mathematical – then the total number of protons, less the total number of electrons, will not change. In fact in this particular model the total number of protons will not change, nor the number of electrons. But we are concentrating now on the charge. The contribution of the protons is positive and that of the electrons negative, and if these objects are never created or destroyed alone then the total charge will be conserved. I want to list as I go on the number of properties that conserve quantities, and I will start with charge (fig. 14). Against the question whether charge is conserved I write 'yes'.

This theoretical interpretation is very simple, but it was later discovered that electrons and protons are not permanent; for example, a particle called the neutron can disintegrate into a proton and an electron – plus something else

	Charge	Baryon No.	Strangeness	Energy	Angular Momentum
Conserved (locally)	Yes	Yes	Nearly	Yes	Yes
Comes in Units	Yes	Yes	Yes	No	Yes
Source of a field	Yes	?	?	Yes	

NB This is the completed table which Professor Feynman added to throughout his lecture.

Figure 14

which we will come to. But the neutron, it turns out, is electrically neutral. So although protons are not permanent, nor are electrons permanent, in the sense that they can be created from a neutron, the charge still checks out; starting before, we had zero charge, and afterwards we had plus one and minus one which when added together become zero charge.

An example of a similar fact is that there exists another particle, besides the proton, which is positively charged. It is called a positron, which is a kind of image of an electron. It is just like the electron in most respects, except that it has the opposite sign of charge, and, more important, it is called an anti-particle because when it meets with an electron the two of them can annihilate each other and disintegrate, and nothing but light comes out. So electrons are not permanent even by themselves. An electron plus a positron will just make light. Actually the 'light' is invisible to the eye; it is gamma rays; but this is the same thing for a physicist, only the wavelength is different. So a particle and its anti-particle can annihilate. The light has no electric charge, but we remove one positive and one negative charge, so we have not changed the total charge. The theory of conservation of charge is therefore slightly more complicated but still very unmathematical. You simply add together the number of positrons you have and the number of protons, take away the number of electrons – there are additional particles you have to check, for example antiprotons which contribute negatively, pi-plus mesons which are positive, in fact each fundamental particle in nature has a charge (possibly zero). All we have to do is add up the total number, and whatever happens in any reaction the total amount of charge on one side has to balance with the amount on the other side.

That is one aspect of the conservation of charge. Now comes an interesting question. Is it sufficient to say only that charge is conserved, or do we have to say more? If charge were conserved because it was a real particle which moved around it would have a very special property. The total amount of charge in a box might stay the same in two ways. It may be that the charge moves from one place to another within the box. But another possibility is that the charge in one place disappears, and simultaneously charge arises in another place, instantaneously related, and in such a manner that the total charge is never changing. This second possibility for the conservation is of a different kind from the first, in which if a charge disappears in one place and turns up in another something has to travel through the space in between. The second form of charge conservation is called local charge conservation, and is far more detailed than the simple remark that the total charge does not change. So you see we are improving our law, if it is true that charge is locally conserved. In fact it is true. I have tried to show you from time to time some of the possibilities of reasoning, of interconnecting one idea with another, and I would now like to describe to you an argument, fundamentally due to Einstein, which indicates that if anything is conserved – and in this case I apply it to charge – it must be conserved locally. This argument relies on one thing, that if two fellows are passing each other in space ships, the question of which guy is doing the moving and which one standing still cannot be resolved by any experiment. That is called the principle of relativity, that uniform motion in a straight line is relative, and that we can look at any phenomenon from either point of view and cannot say which one is standing still and which one is moving.

Suppose I have two space ships, A and B (fig. 15). I am

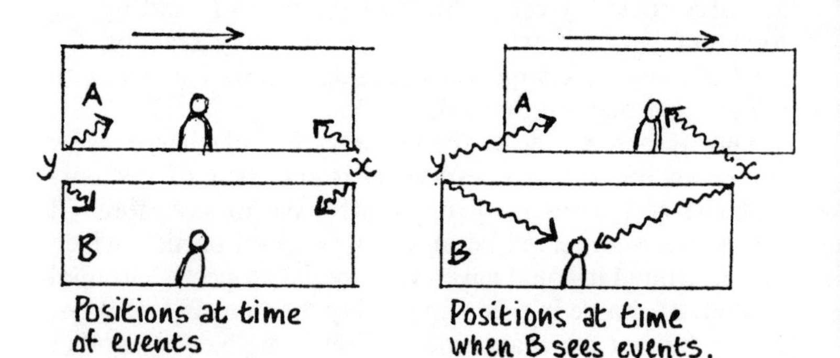

Figure 15

going to take the point of view that A is the one that is moving past B. Remember that is just an opinion, you can also look it at the other way and you will get the same phenomena of nature. Now suppose that the man who is standing still wants to argue whether or not he has seen a charge at one end of his ship disappear and a charge at the other end appear at the same time. In order to make sure it is the same time he cannot sit in the front of the ship, because he will see one before he sees the other because of the travel time of light; so let us suppose that he is very careful and sits dead centre in the middle of the ship. We have another man doing the same kind of observation in the other ship. Now a lightning bolt strikes, and charge is created at point x, and at the same instant at point y at the other end of the ship the charge is annihilated, it disappears. At the same instant, note, and perfectly consistent with our idea that charge is conserved. If we lose one electron in one place we get another elsewhere, but nothing passes in between. Let us suppose that when the charge disappears there is a flash, and when it is created there is a flash, so that we can see what happens. B says they both happen at the same time, since he knows he is in the middle of the ship and the light from the bolt which creates x reaches him at the same time as the light from the flash of disappearance at y. Then B will say, 'Yes, when one disappeared the other was created'. But what happens to our friend in the other ship? He says, 'No, you are wrong my friend. I saw x created before y'. This is because he is moving towards x, so the light from x will have a shorter distance to travel than the light from y, since he is moving away from y. He could say, 'No, x was created first and then y disappeared, so for a short time after x was created and before y disappeared I got some charge. That is not the conservation of charge. It is against the law'. But the first fellow says, 'Yes, but you are moving'. Then he says, 'How do you know? I think you are moving', and so on. If we are unable, by any experiment, to see a difference in the physical laws whether we are moving or not, then if the conservation of charge were not local only a certain kind of man would see it work right, namely the guy who is standing still, in an absolute sense. But such a thing is impossible according to Einstein's relativity principle, and therefore it is impossible to have non-local conservation of charge. The locality of the conservation of charge is consonant with the theory of relativity, and it turns out that this is true of all the conservation laws. You can appreciate that if anything is conserved the same principle applies.

There is another interesting thing about charge, a very strange thing for which we have no real explanation today. It has nothing to do with the conservation law and is independent of it. Charge always comes in units. When we have a charged particle it has one charge or two charges, or minus one or minus two. Returning to our table, although this has nothing to do with the conservation of charge, I must write down that the thing that is conserved comes in units. It is very nice that it comes in units, because that makes the theory of conservation of charge very easy to understand. It is just a *thing* we can count, which goes from place to place. Finally it turns out technically that the total charge of a thing is easy to determine electrically because the charge has a very important characteristic; it is the source of the electric and magnetic field. Charge is a measure of the interaction of an object with electricity, with an electric field. So another item which we should add to the list is that charge is the source of a field; in other words, electricity is related to charge. Thus the particular quantity which is conserved here has two other aspects which are not connected with the conservation directly, but are interesting anyway. One is that it comes in units, and the other that it is the source of a field.

There are many conservation laws, and I will give some more examples of laws of the same type as the conservation of charge, in the sense that it is merely a matter of counting. There is a conservation law called the conservation of baryons. A neutron can go into a proton. If we count each of these as one unit, or baryon, then we do not lose the number of baryons. The neutron carries one baryonic charge unit, or represents one baryon, a proton represents one baryon – all we are doing is counting and making big words! – so if the reaction I am speaking of occurs, in which a neutron decays into a proton, an electron and an anti-neutrino, the total number of baryons does not change. However there are other reactions in nature. A proton plus a proton can produce a great variety of strange objects, for example a lambda, a proton and a K plus. Lambda and K plus are names for peculiar particles.

(easy) $p + p \rightarrow \lambda + p + K^+$

In this reaction we know we put two baryons in, but we see only one come out, so possibly either lambda or K^+ has a baryon. If we study the lambda later we discover that very slowly it disintegrates into a proton and a pi, and ultimately the pi disintegrates into electrons and what-not.

$$(\text{slow})\ \lambda \rightarrow P + \pi$$

What we have here is the baryon coming out again in the proton, so we think the lambda has a baryon number of 1, but the K^+ does not, the K^+ has zero.

On our chart of conservation laws (fig. 14), then, we have charge and now we have a similar situation with baryons, with a special rule that the baryon number is the number of protons, plus the number of neutrons, plus the number of lambdas, minus the number of anti-protons, minus the number of anti-neutrons, and so on; it is just a counting proposition. It is conserved, it comes in units, and nobody knows but everybody wants to think, by analogy, that it is the source of a field. The reason we make these tables is that we are trying to guess at the laws of nuclear interaction, and this is one of the quick ways of guessing at nature. If charge is the source of a field, and baryon does the same things in other respects it ought to be the source of a field too. Too bad that so far it does not seem to be, it is possible, but we do not know enough to be sure.

There are one or two more of these counting propositions, for example Lepton numbers, and so on, but the idea is the same as with baryons. There is one, however, which is slightly different. There are in nature among these strange particles characteristic rates of reaction, some of which are very fast and easy, and others which are very slow and hard. I do not mean easy and hard in a technical sense, in actually doing the experiment. It concerns the rates at which the reactions occur when the particles are present. There is a clear distinction between the two kinds of reaction which I have mentioned above, the decay of a pair of protons, and the much slower decay of the lambda. It turns out that if you take only the fast and easy reactions there is one more counting law, in which the lambda gets a minus 1, and the K plus gets a plus 1, and the proton gets zero. This is called the strangeness number, or hyperon charge, and it appears that the rule that it is conserved is right for every easy reaction, but wrong for the slow reactions. On our chart (fig. 14) we must therefore add the conservation law called the conservation of strangeness, or the conservation of hyperon number, which is nearly right. This is very peculiar; we see why this quantity has been called strangeness. It is nearly true that it is conserved, and true that it comes in units. In trying to understand the strong interactions which are involved in nuclear forces, the fact that in strong interactions the thing is conserved has made people propose that for strong interactions it is also the source of a field, but again we do not know. I bring these matters up to show you how conservation laws can be used to guess new laws.

There are other conservation laws that have been proposed from time to time, of the same nature as counting. For example, chemists once thought that no matter what happened the number of sodium atoms stayed the same. But sodium atoms are not permanent. It is possible to transmute atoms from one element to another so that the original element has completely disappeared. Another law which was for a while believed to be true was that the total mass of an object stays the same. This depends on how you define mass, and whether you get mixed up with energy. The mass conservation law is contained in the next one which I am going to discuss, the law of conservation of energy. Of all the conservation laws, that dealing with energy is the most difficult and abstract, and yet the most useful. It is more difficult to understand than those I have described so far, because in the case of charge, and the others, the mechanism is clear, it is more or less the conservation of objects. This is not absolutely the case, because of the problem that we get new things from old things, but it is really a matter of simply counting.

The conservation of energy is a little more difficult, because this time we have a number which is not changed in time, but this number does not represent any particular thing. I would like to make a kind of silly analogy to explain a little about it.

I want you to imagine that a mother has a child whom she leaves alone in a room with 28 absolutely indestructible blocks. The child plays with the blocks all day, and when the mother comes back she discovers that there are indeed 28 blocks; she checks all the time the conservation of blocks! This goes on for a few days, and then one day when she comes in there are only 27 blocks. However, she finds one block lying outside the window, the child had thrown it out. The first thing you must appreciate about conservation laws is that you must watch that the stuff you are trying to check does not go out through the wall. The same thing could happen the other way, if a boy came in to play with the child, bringing some blocks with him. Obviously these are matters you have to consider when you talk about conservation laws. Suppose one day when the mother comes to count the blocks she finds that there are only 25 blocks, but suspects that the child has hidden the other three blocks in a little toy box. So she says, 'I am going to open the box'. 'No,' he says, 'you cannot open the box.' Being a very clever mother she would say, 'I know that when the box is empty it weighs 16 ounces, and each block weighs 3 ounces, so what I am going to do is to weigh the box'. So, totalling up the number of blocks, she would get –

$$\text{No. of blocks seen} + \frac{\text{Weight of box} - 16\text{ oz.}}{3\text{ oz.}}$$

and that adds up to 28. This works all right for a while, and then one day the sum does not check up properly. However, she notices that the dirty water in the sink is changing its level. She knows that the water is 6 inches deep when there is no block in it, and that it would rise $\frac{1}{4}$ inch if a block was in the water, so she adds another term, and now she has –

$$\text{No. of blocks seen} + \frac{\text{Weight of box} - 16\text{oz.}}{3\text{oz.}} + \frac{\text{Ht. of Water} - 6\text{in.}}{\frac{1}{4}\text{in.}}$$

and once again it adds up to 28. As the boy becomes more ingenious, and the mother continues to be equally ingenious, more and more terms must be added, all of which represent blocks, but from the mathematical standpoint are abstract calculations, because the blocks are not seen.

Now I would like to draw my analogy, and tell you what is common between this and the conservation of energy, and what is different. First suppose that in all of the situations you never saw any blocks. The term 'No. of blocks seen' is never included. Then the mother would always be calculating a whole lot of terms like 'blocks in the box', 'blocks in the water', and so on. With energy there is this difference, that there are no blocks, so far as we can tell. Also, unlike the case of the blocks, for energy the numbers that come out are not integers. I suppose it might happen to the poor mother that when she calculates one term it comes out $6\frac{1}{8}$ blocks, and when she calculates another it comes out $\frac{7}{8}$ of a block, and the others give 21, which still totals 28. That is how it looks with energy.

What we have discovered about energy is that we have a scheme with a sequence of rules. From each different set of rules we can calculate a number for each different kind of energy. When we add all the numbers together, from all the different forms of energy, it always gives the same total. But as far as we know there are no real units, no little ball-bearings. It is abstract, purely mathematical, that there is a number such that whenever you calculate it it does not change. I cannot interpret it any better than that.

This energy has all kinds of forms, analogous to the blocks in the box, blocks in the water, and so on. There is energy due to motion called kinetic energy, energy due to gravitational interaction (gravitational potential energy, it is called), thermal energy, electrical energy, light energy, elastic energy in springs and so on, chemical energy, nuclear energy – and there is also an energy that a particle has from its mere existence, an energy that depends directly on its mass. The last is the contribution of Einstein, as you undoubtedly know. $E = mc^2$ is the famous equation of the law I am talking about.

Although I have mentioned a large number of energies, I would like to explain that we are not completely ignorant about this, and we do understand the relationship of some of them to others. For instance, what we call thermal energy is to a large extent merely the kinetic energy of the motion of the particles inside an object. Elastic energy and chemical energy both have the same origin, namely the forces between the atoms. When the atoms rearrange themselves in a new pattern some energy is changed, and if that quantity changes it means that some other quantity also has to change. For example, if you are burning something the chemical energy changes, and you find heat where you did not have heat before, because it all has to add up right. Elastic energy and chemical energy are both interactions of atoms, and we now understand these interactions to be a combination of two things, one electrical energy and the other kinetic energy again, only this time the formula for it is quantum mechanical. Light energy is nothing but electrical energy, because light has now been interpreted as an electric and magnetic wave. Nuclear energy is not represented in terms of the others; at the moment I cannot say more than that it is the result of nuclear forces. I am not just talking here about the energy released. In the uranium nucleus there is a certain amount of energy, and when the thing disintegrates the amount of energy remaining in the nucleus changes, but the total amount of energy in the world does not change, so a lot of heat and stuff is generated in the process, in order to balance up.

This conservation law is very useful in many technical ways. I will give you some very simple examples to show how, knowing the law of conservation of energy and the formulae for calculating energy, we can understand other laws. In other words many other laws are not independent, but are simply secret ways of talking about the conservation of energy. The simplest is the law of the lever (fig. 16).

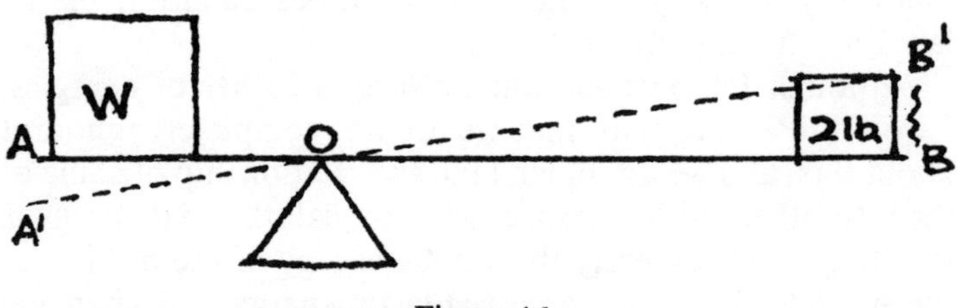

Figure 16

We have a lever on a pivot. The length of one arm is 1 foot and the other 4 feet. First I must give the law for gravity energy, which is that if you have a number of weights, you take the weight of each and multiply it by its height above the ground, add this together for all the weights, and that gives the total of gravity energy. Suppose I have a 2 lb weight on the long arm, and an unknown mystic weight on the other side – X is always the unknown, so let us call it W to make it seem that we have advanced above the usual! Now the question is, how much must W be so that it just balances and swings quietly back and forth without any trouble? If it swings quietly back and forth, that means that the energy is the same whether the balance is parallel to the ground or tilted so that the 2 lb weight is, say, 1 inch above the ground. If the energy is the same then it does not care much which way, and it does not fall over. If the 2 lb weight goes up 1 inch how far down does W go? From the diagram you can see (fig. 3) that if AO is 1 foot and OB is 4 feet, then when BB′ is 1 inch AA′ will be $\frac{1}{4}$ inch. Now apply the law for gravity energy. Before anything happened all the heights were zero, so the total energy was zero. After the move has happened to get the gravity energy we multiply the weight 2 lb by the height 1 inch and add it to the unknown weight W times the height – $\frac{1}{4}$ inch. The sum of this must give the same energy as before – zero. So –

$$2 - \frac{W}{4} = 0, \text{ so } W \text{ must be } 8$$

This is one way we can understand the easy law, which you already knew of course, the law of the lever. But it is interesting that not only this but hundreds of other physical laws can be closely related to various forms of energy. I showed you this example only to illustrate how useful it is.

The only trouble is, of course, that in practice it does not really work because of friction in the fulcrum. If I have something moving, for example a ball rolling along at a constant height, then it will stop on account of friction. What happened to the kinetic energy of the ball? The answer is that the energy of the motion of the ball has gone into the energy of the jiggling of the atoms in the floor and in the ball. The world that we see on a large scale looks like a nice round ball when we polish it, but it is really quite complicated when looked at on a little scale; billions of tiny atoms, with all kinds of irregular shapes. It is like a very rough boulder when looked at finely enough, because it is made out of these little balls. The floor is the same, a bumpy business made out of balls. When you roll this monster boulder over the magnified floor you can see that the little atoms are going to go snap-jiggle, snap-jiggle. After the thing has rolled across, the ones that are left behind are still shaking a little from the pushing and snapping that they went through; so there is left in the floor a jiggling motion, or thermal energy. At first it appears as if the law of conservation is false, but energy has the tendency to hide from us and we need thermometers and other instruments to make sure that it is still there. We find that energy is conserved no matter how complex the process, even when we do not know the detailed laws.

The first demonstration of the law of conservation of

energy was not by a physicist but by a medical man. He demonstrated with rats. If you burn food you can find out how much heat is generated. If you then feed the same amount of food to rats it is converted, with oxygen, into carbon dioxide, in the same way as in burning. When you measure the energy in each case you find out that living creatures do exactly the same as non-living creatures. The law for conservation of energy is as true for life as for other phenomena. Incidentally, it is interesting that every law or principle that we know for 'dead' things, and that we can test on the great phenomenon of life, works just as well there. There is no evidence yet that what goes on in living creatures is necessarily different, so far as the physical laws are concerned, from what goes on in non-living things, although the living things may be much more complicated.

The amount of energy in food, which will tell you how much heat, mechanical work, etc., it can generate, is measured in calories. When you hear of calories you are not eating something called calories, that is simply the measure of the amount of heat energy that is in the food. Physicists sometimes feel so superior and smart that other people would like to catch them out once on something. I will give you something to get them on. They should be utterly ashamed of the way they take energy and measure it in a host of different ways, with different names. It is absurd that energy can be measured in calories, in ergs, in electron volts, in foot pounds, in B.T.U.s, in horsepower hours, in kilowatt hours – all measuring exactly the same thing. It is like having money in dollars, pounds, and so on; but unlike the economic situation where the ratio can change, these dopey things are in absolutely guaranteed proportion. If anything is analogous, it is like shillings and pounds – there are always 20 shillings to a pound. But one complication that the physicist allows is that instead of having a number like 20 he has irrational ratios like 1·6183178 shillings to a pound. You would think that at least the more modern high-class theoretical physicists would use a common unit, but you find papers with degrees Kelvin for measuring energy, megacycles, and now inverse Fermis, the latest invention. For those who want some proof that physicists are human, the proof is in the idiocy of all the different units which they use for measuring energy.

There are a number of interesting phenomena in nature which present us with curious problems concerning energy. There has been a recent discovery of things called quasars, which are enormously far away, and they radiate so much energy in the form of light and radio waves that the question is where does it come from? If the conservation of energy is right, the condition of the quasar after it has radiated this enormous amount of energy must be different from its condition before. The question is, is it coming from gravitation energy – is the thing collapsed gravitationally, in a different condition gravitationally? Or is this big emission coming from nuclear energy? Nobody knows. You might propose that perhaps the law of conservation of energy is not right. Well, when a thing is investigated as incompletely as the quasar – quasars are so distant that the astronomers cannot see them too easily – then if such a thing seems to conflict with the fundamental laws, it very rarely is that the fundamental laws are wrong, it usually is just that the details are unknown.

Another interesting example of the use of the law of conservation of energy is in the reaction when a neutron disintegrates into a proton, an electron, and an anti-neutrino. It was first thought that a neutron turned into a proton plus an electron. But the energy of all the particles could be measured, and a proton and an electron together did not add up to a neutron. Two possibilities existed. It might have been that the law of energy conservation was not right; in fact it was proposed by Bohr* for a while that perhaps the conservation law worked only statistically, on the average. But it turns out now that the other possibility is the correct one, that the fact that the energy does not check out is because there is something else coming out, something

***Niels Bohr, Danish physicist.**

which we now call an anti-neutrino. The anti-neutrino which comes out takes up the energy. You might say that the only reason for the anti-neutrino is to make the conservation of energy right. But it makes a lot of other things right, like the conservation of momentum and other conservation laws, and very recently it has been directly demonstrated that such neutrinos do indeed exist.

This example illustrates a point. How is it possible that we can extend our laws into regions we are not sure about? Why are we so confident that, because we have checked the energy conservation here, when we get a new phenomenon we can say it has to satisfy the law of conservation of energy? Every once in a while you read in the papeı that physicists have discovered that one of their favourite laws is wrong. Is it then a mistake to say that a law is true in a region where you have not yet looked? If you will never say that a law is true in a region where you have not already looked you do not know anything. If the only laws that you find are those which you have just finished observing then you can never make any predictions. Yet the only utility of science is to go on and to try to make guesses. So what we always do is to stick our necks out, and in the case of energy the most likely thing is that it is conserved in other places.

Of course this means that science is uncertain; the moment that you make a proposition about a region of experience that you have not directly seen then you must be uncertain. But we always must make statements about the regions that we have not seen, or the whole business is no use. For instance, the mass of an object changes when it moves, because of the conservation of energy. Because of the relation of mass and energy the energy associated with the motion appears as an extra mass, so things get heavier when they move. Newton believed that this was not the case, and that the masses stayed constant. When it was discovered that the Newtonian idea was false everyone kept saying what a terrible thing it was that physicists had found out that they were wrong. Why did they think they were right? The effect is very small, and only shows when you get near the speed of light. If you spin a top it weighs the same as if you do not spin it, to within a very very fine fraction. Should they then have said, 'If you do not move any faster than so-and-so, then the mass does not change'? That would then be certain. No, because if the experiment happened to have been done only with tops of wood, copper and steel, they would have had to say 'Tops made out of copper, wood and steel, when not moving any faster than so and so . . .'. You see, we do not know all the conditions that we need for an experiment. It is not known whether a radioactive top would have a mass that is conserved. So we have to make guesses in order to give any utility at all to science. In order to avoid simply describing experiments that have been done, we have to propose laws beyond their observed range. There is nothing wrong with that, despite the fact that it makes science uncertain. If you thought before that science was certain – well, that is just an error on your part.

To return then, to our list of conservation laws (fig. 14), we can add energy. It is conserved perfectly, as far as we know. It does not come in units. Now the question is, is it the source of a field? The answer is yes. Einstein understood gravitation as being generated by energy. Energy and mass are equivalent, and so Newton's interpretation that the mass is what produces gravity has been modified to the statement that the energy produces the gravity.

There are other laws similar to the conservation of energy, in the sense that they are numbers. One of them is momentum. If you take all the masses of an object, multiply them by the velocities, and add them all together, the sum is the momentum of the particles; and the total amount of momentum is conserved. Energy and momentum are now understood to be very closely related, so I have put them in the same column of our table.

Another example of a conserved quantity is angular momentum, an item which we discussed before. The angular momentum is the area generated per second by objects moving about. For example, if we have a moving object, and we take any centre whatsoever, then the speed at which the area (fig. 17) swept out by a line from centre to object,

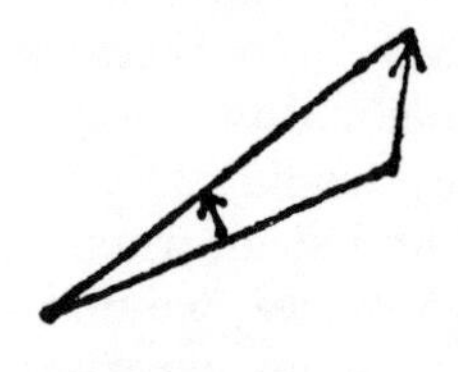

Figure 17

increases, multiplied by the mass of the object, and added together for all the objects, is called the angular momentum. And that quantity does not change. So we have conservation of angular momentum. Incidentally, at first sight, if you know too much physics, you might think that the angular momentum is not conserved. Like the energy it appears in different forms. Although most people think it only appears in motion it does appear in other forms, as I will illustrate. If you have a wire, and move a magnet up into it, increasing the magnetic field through the flux through the wire, there will be an electric current – that is how electric generators work. Imagine that instead of a wire I have a disc, on which there are electric charges analogous to the electrons in the wire (fig. 18). Now I bring a magnet dead centre along the

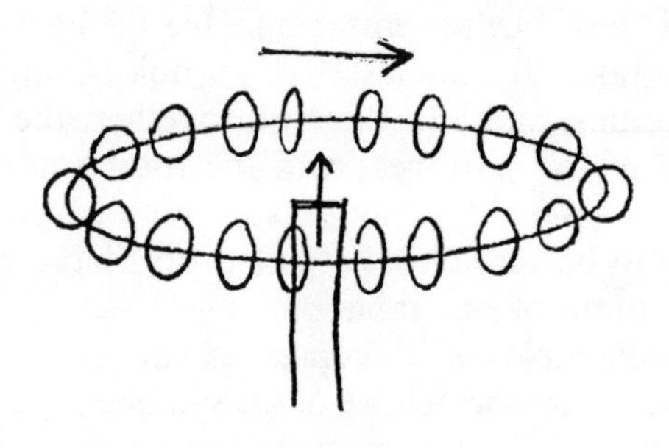

Figure 18

axis from far away, very rapidly up to the disc, so that now there is a flux change. Then, just as in the wire, the charges will start to go around, and if the disc were on a wheel it would be spinning by the time I had brought the magnet up. That does not look like conservation of angular momentum, because when the magnet is away from the disc nothing is turning, and when they are close together it is spinning. We have got turning for nothing, and that is against the rules. 'Oh yes,' you say, 'I know, there must be some other kind of interaction that makes the magnet spin the opposite way.' That is not the case. There is no electrical force on the magnet tending to twist it the opposite way. The explanation is that angular momentum appears in two forms: one of them is angular momentum of motion, and the other is angular momentum in electric and magnetic fields. There is angular momentum in the field around the magnet, although it does not appear as motion, and this has the opposite sign to the spin. If we take the opposite case it is even clearer (fig. 19).

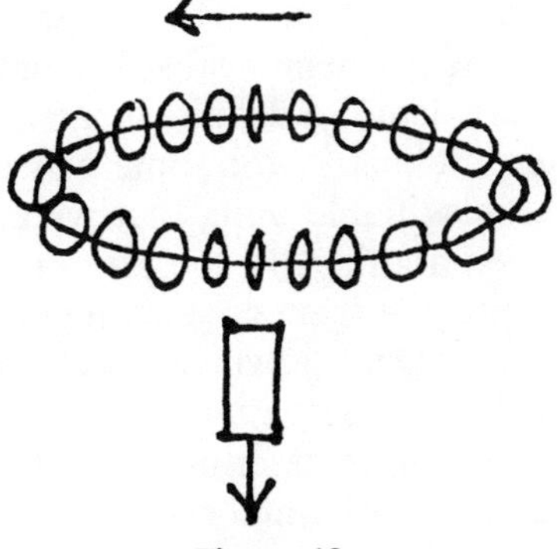

Figure 19

If we have just the particles, and the magnet, close together, and everything is standing still, I say there is angular momentum in the field, a hidden form of angular momentum which does not appear as actual rotation. When you pull the magnet down and take the instrument apart, then all the fields separate and the angular momentum now has to appear and the disc will start to spin. The law that makes it spin is the law of induction of electricity.

Whether angular momentum comes in units is very difficult for me to answer. At first sight it appears that it is absolutely impossible that angular momentum comes in units, because angular momentum depends upon the direction at which you project the picture. You are looking at an area change, and obviously this will be different depending on whether it is looked at from an angle, or straight on. If angular momentum came in units, and say you looked at something and it showed 8 units, then if you looked at it from a very slightly different angle, the number of units would be very slightly different, perhaps a tiny bit less than 8. But 7 is not a little bit less than 8; it is a definite amount less than eight. So it cannot possibly come in units. However this proof is evaded by the subtleties and peculiarities of quantum mechanics, and if we measure the angular momentum about any axis, amazingly enough it is always a number of units. It is not the kind of unit, like an electric charge, that you can count. The angular momentum does come in units in the mathematical sense that the number we get in any measurement is a definite integer times a unit. But we cannot interpret this in the same way as with units of electric charge, imaginable units that we can count – one, then another, then another. In the case of angular momentum we cannot imagine them as separate units, but it comes out always as an integer . . . which is very peculiar.

There are other conservation laws. They are not as interesting as those I have described, and do not deal exactly with the conservation of numbers. Suppose we had some kind of device with particles moving with a certain definite symmetry, and suppose their movements were bilaterally symmetrical (fig. 20). Then, following the laws of physics, with all the movements and collisions, you could expect, and rightly, that if you look at the same picture later on it will still be bilaterally symmetrical. So there is a kind of conservation, the conservation of the symmetry character. This should be in the table, but it is not like a number that you

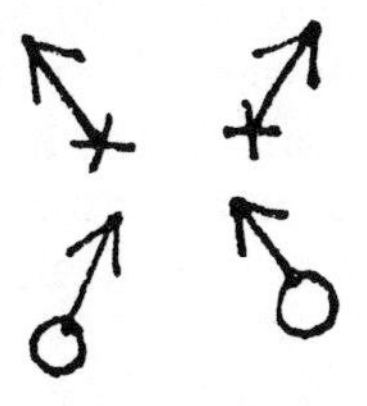

Figure 20

measure, and we will discuss it in much more detail in the next lecture. The reason this is not very interesting in classical physics is because the times when there are such nicely symmetrical initial conditions are very rare, and it is therefore a not very important or practical conservation law. But in quantum mechanics, when we deal with very simple systems like atoms, their internal constitution often has a kind of symmetry, like bilateral symmetry, and then the symmetry character is maintained. This is therefore an important law for understanding quantum phenomena.

One interesting question is whether there is a deeper basis for these conservation laws, or whether we have to take them as they are. I will discuss that question in the next lecture, but there is one point I should like to make now. In discussing these ideas on a popular level, there seem to be a lot of unrelated concepts; but with a more profound understanding of the various principles there appear deep interconnections between the concepts, each one implying others in some way. One example is the relation between relativity and the necessity for local conservation. If I had stated this without a demonstration, it might appear to be some kind of miracle that if you cannot tell how fast you are moving this implies that if something is conserved it must be done not by jumping from one place to another.

At this point I would like to indicate how the conservation of angular momentum, the conservation of momentum, and a few other things are to some extent related. The conservation of angular momentum has to do with the area swept by particles moving. If you have a lot of particles (fig. 21), and take your centre (x) very far away, then the distances are almost the same for every object. In this case the only thing that counts in the area sweeping, or in the conservation of angular momentum, is the component of motion, which in figure 21 is vertical. What we discover then

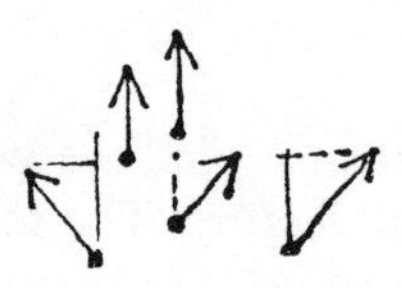

Figure 21

is that the total of the masses, each multiplied by its velocity vertically, must be a constant, because the angular momentum is a constant about any point, and if the chosen point is far enough away only the masses and velocities are relevant. In this way the conservation of angular momentum implies the conservation of momentum. This in turn implies something else, the conservation of another item which is so closely connected that I did not bother to put it in the table. This is a principle about the centre of gravity (fig. 22).

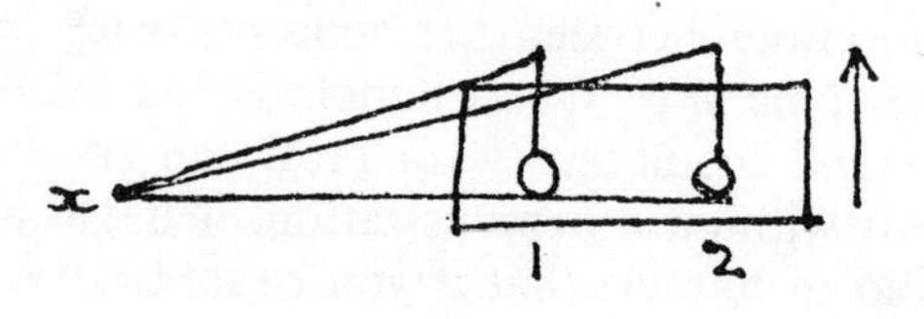

Figure 22

A mass, in a box, cannot just disappear from one position and move over to another position all by itself. That is nothing to do with conservation of the mass; you still have the mass, just moved from one place to another. Charge could do this, but not a mass. Let me explain why. The laws of physics are not affected by motion, so we can suppose that this box is drifting slowly upwards. Now we take the angular momentum from a point not far away, x. As the box is drifting upwards, if the mass is lying quiet in the box, at position 1, it will be producing an area at a given rate. After the mass has moved over to position 2, the area will be increasing at a greater rate, because although the altitude will be the same because the box is still drifting upwards, the distance from x to the mass has increased. By the conservation of angular momentum you cannot change the rate at which the area is changing, and therefore you simply cannot move one mass from one place to another unless you push on something else to balance up the angular momentum. That is the reason why rockets in empty space cannot go . . . but they do go. If you figure it out with a lot of masses, then if you move one forward you must move others back, so that the total motion back and forward of all the masses is nothing. This is how a rocket works. At first it is standing still, say, in empty space, and then it shoots some gas out of the back, and the rocket goes forward. The point is that of all the stuff in the world, the centre of mass, the average of all the mass, is still right where it was before. The interesting part has moved on, and an uninteresting part that we do not care about has moved back. There is no theorem that says that the interesting things in the world are conserved – only the total of everything.

Discovering the laws of physics is like trying to put together the pieces of a jigsaw puzzle. We have all these different pieces, and today they are proliferating rapidly. Many of them are lying about and cannot be fitted with the other ones. How do we know that they belong together? How do we know that they are really all part of one as yet incomplete picture? We are not sure, and it worries us to some extent, but we get encouragement from the common characteristics of several pieces. They all show blue sky, or they are all made out of the same kind of wood. All the various physical laws obey the same conservation principles.

4

Symmetry in Physical Law

Symmetry seems to be absolutely fascinating to the human mind. We like to look at symmetrical things in nature, such as perfectly symmetrical spheres like planets and the sun, or symmetrical crystals like snowflakes, or flowers which are nearly symmetrical. However, it is not the symmetry of the objects in nature that I want to discuss here; it is rather the symmetry of the physical laws themselves. It is easy to understand how an object can be symmetrical, but how can a physical law have a symmetry? Of course it cannot, but physicists delight themselves by using ordinary words for something else. In this case they have a feeling about the physical laws which is very close to the feeling of symmetry of objects, and they call it the symmetry of the laws. That is what I am going to discuss.

What is symmetry? If you look at me I am symmetrical, right and left – apparently externally, at least. A vase can be symmetrical in the same way or in other ways. How can you define it? The fact that I am left and right symmetric means that if you put everything that is on one side on the other side, and vice versa – if you just exchange the two sides – I shall look exactly the same. A square has a symmetry of a special kind, because if I turn it around through 90 degrees it still looks exactly the same. Professor Weyl,* the mathematician, gave an excellent definition of symmetry, which is that a thing is symmetrical if there is something that you can do to it so that after you have finished doing it it looks the same as it did before. That is the sense in which we say that the laws of physics are symmetrical; that there are

*Hermann Weyl, 1885–1955, German mathematician.

things we can do to the physical laws, or to our way of representing the physical laws, which make no difference, and leave everything unchanged in its effects. It is this aspect of physical laws that is going to concern us in this lecture.

The simplest example of this kind of symmetry – you will see that it is not the same as you might have thought, left and right symmetric, or anything like that – is a symmetry called translation in space. This has the following meaning: if you build any kind of apparatus, or do any kind of experiment with some things, and then go and build the same apparatus to do the same kind of experiment, with similar things but put them here instead of there, merely translated from one place to another in space, then the same thing will happen in the translated experiment as would have happened in the original experiment. It is not true here actually. If I actually built such an apparatus, and then displaced it 20 feet to the left of where I am now it would get into the wall, and there would be difficulties. It is necessary, in defining this idea, to take into account everything that might affect the situation, so that when you move the thing you move everything. For example, if the system involved a pendulum, and I moved it 20,000 miles to the right, it would not work properly any more because the pendulum involves the attraction of the earth. However, if I imagine that I move the earth as well as the equipment then it would behave in the same way. The problem in this situation is that you must translate everything which may have any influence on the situation. That sounds a little dopey, because it sounds as if you can just translate an experiment, and if it does not work you can just presume that you did not translate enough stuff – so you are bound to win. Actually this is not so, because it is not self-evident that you are bound to win. The remarkable thing about nature is that it is possible to translate enough stuff so that it does behave the same way. That is a positive statement.

I would like to illustrate that such a thing is true. Let us take as an example the law of gravitation, which says that the force between objects varies inversely as the square of the distance between them; and I would remind you that a thing responds to a force by changing its velocity, with time, in the direction of the force. If I have a pair of objects, like a planet going around a sun, and I move the whole pair over, then the distance between the objects of course does not change, and so the forces do not change. Further, when they are in the moved-over situation they will go at the same speed, and all the changes will remain in proportion and everything go around in the two systems in exactly the same way. The fact that the law says 'the distance between two objects', rather than some absolute distance from the central eye of the universe, means that the laws are translatable in space.

That, then, is the first symmetry – translation in space. The next one could be called translation in time, but, better, let us say that delay in time makes no difference. We start a planet going around the sun in a certain direction; if we could start it all over again, two hours later, or two years later, with another beginning, but starting with the planet and the sun going in exactly the same way, then it would behave in exactly the same way, because again the law of gravitation talks about the velocity, and never about the absolute time when you were supposed to start measuring things. In this particular example, in fact, we are not really sure. When we discussed gravitation, we talked about the possibility that the force of gravity changed with time. This would mean that translation in time is not a valid proposition, because if the constant of gravitation will be weaker a billion years hence than it is now, then it is not true that the motion would be exactly the same for our experimental sun and planet a billion years from now as it is now. As far as we know today (I have only discussed the laws as we know them today. – I only wish I could discuss the laws as we shall know them tomorrow!) as far as we know, a delay in time makes no difference.

We know that in one respect this is not really true. It is true for what we now call physical laws; but one of the facts of the world (which may be very different) is that it looks as if the universe had a definite time of beginning, and that everything is exploding apart. You might call that a condition of geography, analogous to the situation that when I translate in space I must translate everything. In the same sense you might say that the laws for time are the same and we must move the expansion of the universe with everything else. We could have made another analysis in which we started the universe later; but we do not start the universe, and we have no control over the situation and no way to define that idea experimentally. Therefore as far as science is concerned there really is no way to tell. The fact of the matter is that the conditions of the world appear to be changing in time, the galaxies separating from one another, so if you were to awake in some science-fiction story at an unknown time, by measuring the average distances to the galaxies you could tell when it was. That means that the world will not look the same if delayed in time.

Now it is conventional today to separate the physical laws, which tell how things will move if you start them in a given condition, from the statement of how the world actually began, because we know so little about that. It is usually considered that astronomical history, or cosmological history, is a little different from physical law. Yet if put to a test of how to define the difference I would be hard pressed. The best characteristic of physical law is its universality, and if anything is universal it is the expansion of all the nebulae. I have therefore no way of defining the difference. However, if I restrict myself to disregard the origin of the universe and take only the physical laws that are known, then a delay in time makes no difference.

Let us take some other examples of symmetry laws. One is a rotation in space, a fixed rotation. If I do some experiments with a piece of equipment built in one place, and then take another one (possibly translated so that it does not get in the way) exactly the same, but turned so that all the axes are in a different direction, it will work the same way. Again we have to turn everything that is relevant. If the thing is a grandfather clock, and you turn it horizontal, then the pendulum will just sit up against the wall of the cabinet and not work. But if you turn the earth too (which is happening all the time) the clock still keeps working.

The mathematical description of this possibility of turning is a rather interesting one. To describe what goes on in a situation we use numbers to tell where something is. They are called the co-ordinates of a point, and we sometimes use three numbers, to describe how high the point is above some plane, how far it is in front, say, or behind in negative numbers, and how far to the left. In this case I am not going to worry about up and down because for rotations I just have to use two of these three co-ordinates. Let us call the distance in front of me x, and y can be the distance to the left. Then I can locate any body by telling how far it is in front and how far to the left. Those who come from New York City will know that the street numbers work that way very neatly – or they did until they began to change the name of Sixth Avenue! The mathematical idea about the turning is this: if I locate a point as I have described, by giving its x and y co-ordinates and someone else, facing a different way, locates the same point in the same way, but calculating the x′ and y′ in relation to his own position, then you can see that my x co-ordinate is a mixture of the two co-ordinates calculated by the other man. The connexion of the transformation is that x gets mixed into x′ and y′ and y into y′ and x′. The laws of nature should so be written that if you make such a mixture, and resubstitute in the equations, then the equations will not change their form. That is the way in which the symmetry appears in mathematical form. You write the equations with certain letters, then there is a way of changing the letters from x and y to a different x, x′, and a different y, y′, which is a formula in terms of the old x and y, and the equations look the same, only they have primes all over them. This just means that the other man will see the thing behaving in his apparatus the same way as I see it in mine, which is turned the other way.

I will give another, very interesting, example of a symmetry law. It is a question of uniform velocity in a straight

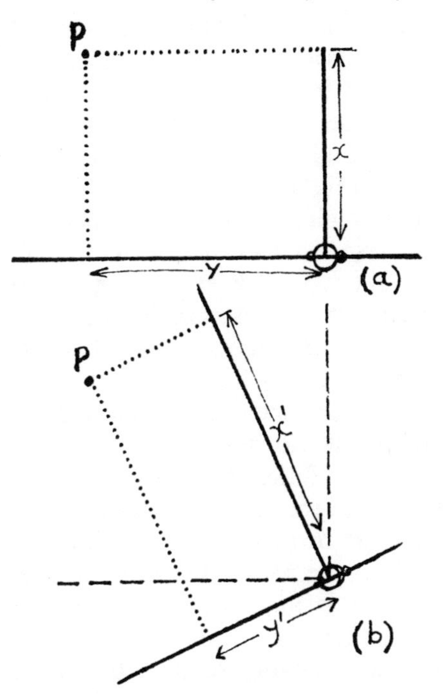

Relation of point P to me is described by two numbers x, y; x is how far P is in front of me and y is how far to the left.

The same point P is described by two new numbers x′, y′ if I am in the same place but simply turned.

Figure 23

line. It is believed that the laws of physics are unchanged under a uniform velocity in a straight line. This is called the principle of relativity. If we have a space ship, and we have a bit of equipment in it that is doing something, and we have another similar equipment down here on the ground, then, if the space ship is going along at a uniform speed, somebody inside, watching what is going on on his apparatus, can see nothing different from the effects I, who am standing still, can see on my apparatus. Of course if he looks outside, or if he bumps into an outside wall, or something like that, that is another matter; but in so far as he is moving at a uniform velocity in a straight line, the laws of physics look the same to him as they do to me. Since that is the case, I cannot say who is moving.

I must emphasize here, before we go any further, that in all of these transformations, and all of these symmetries, we are not talking about moving a whole universe. In the case of time I am saying nothing if I imagine that I move all the times in the whole universe. So also there would be no content in the statement that if I took everything in the whole universe, and moved it over in space, it would behave the same way. The remarkable thing is that if I take a piece of apparatus and move it over, then if I make sure about a lot of conditions, and include enough apparatus, I can get a piece of the world and move it relative to the average of all the rest of the stars, and this still does not make any difference. In the relativity case it means that someone coasting at a uniform velocity in a straight line, relative to the average of the rest of the nebulae, sees no effect. Put another way, it is impossible to determine by any effects from the experiments inside a car, without looking out, whether you are moving relative to all the stars.

This proposition was first stated by Newton. Let us take his law of gravitation. It says that the forces are inversely as the squares of the distances, and that a force produces a change in velocity. Now suppose I have worked out what happens when a planet goes around a fixed sun, and now I want to work out what happens when a planet is going around a drifting sun. Then all of the velocities that I had in the first case are different in the second case; I have to add on a constant velocity. But the law is stated in terms of *changes* in velocity, so that what happens is that the force on the planet with the fixed sun is the same as the force on the planet with the drifting sun, and therefore the changes in velocity of the two planets will also be identical. So any extra velocity I started with on the second planet just keeps on going, and all the changes are accumulated on top of that. The net result of the mathematics is that if you add a constant speed the laws will be exactly the same, so that we cannot, by studying the solar system and the way the planets go around the sun, figure out whether the sun is itself drifting through space. According to Newton's law there is no effect of such a drift through space on the motions of the planets around the sun; so Newton added that 'The motion of bodies among themselves is the same in a space, whether that space is itself at rest relative to the fixed stars, or moving at a uniform velocity in a straight line'.

As time went on, new laws were discovered after Newton, among them the laws of electricity discovered by Maxwell.* One of the consequences of the laws of electricity was that there should be waves, electromagnetic waves – light is an example – which should go at 186,000 miles a second, *flat.* I mean by that 186,000 miles a second, come what may. So then it was easy to tell where rest was, because the law that light goes at 186,000 miles a second is certainly not (at first sight) one which will permit one to move without some effect. It is evident, is it not, that if you are in a space ship going at 100,000 miles a second in some direction, while I am standing still, and I shoot a light beam at 186,000 miles a second through a little hole in your ship, then, as it goes through your ship, since you are going at 100,000 miles per second and the light is going at 186,000, the light is only going to look to you as if it is passing at 86,000 miles a second. But it turns out that if you do this experiment it looks to you as if it is going at 186,000 miles a second past you, and to me as if it is going 186,000 miles a second past me!

The facts of nature are not so easy to understand, and the fact of the experiment was so obviously counter to commonsense, that there are some people who still do not believe the result! But time after time experiments indicated that the speed is 186,000 miles a second no matter how fast you are moving. The question now is how that could be. Einstein

*James Clerk Maxwell, 1831–79. First teacher of experimental physics at Cambridge.

realized, and Poincaré* too, that the only possible way in which a person moving and a person standing still could measure the speed to be the same was that their sense of time and their sense of space are not the same, that the clocks inside the space ship are ticking at a different speed from those on the ground, and so forth. You might say, 'Ah, but if the clock is ticking and I look at the clock in the space ship, then I can see that it is going slow'. No, your brain is going slow too! So by making sure that everything went just so inside the space ship, it was possible to cook up a system by which in the space ship it would look like 186,000 space-ship miles per space-ship second, whereas here it would look like 186,000 my miles per my second. That is a very ingenious thing to be able to do, and it turns out, remarkably enough, to be possible.

I have mentioned already one of the consequences of this principle of relativity, that you cannot tell how fast you are moving in a straight line; you remember in the last lecture the case in which we had two cars, A and B (fig. 24). There was an event, which happened at each end of car B. A man

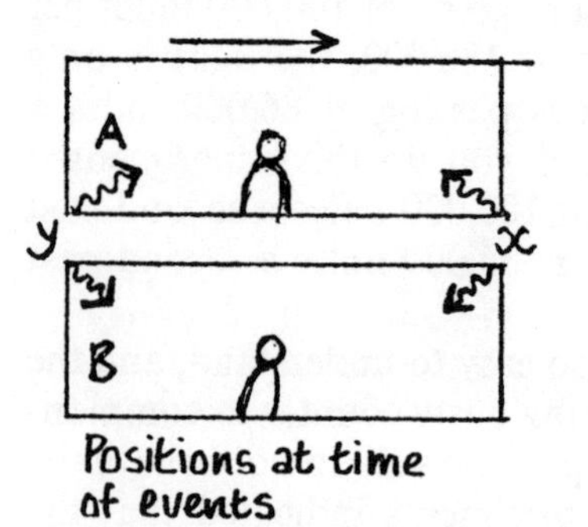

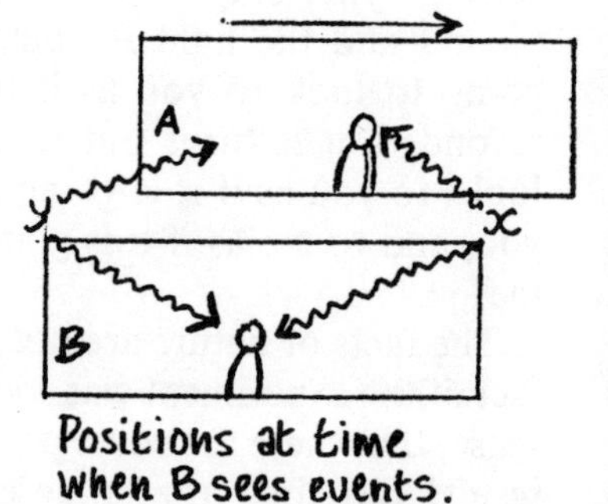

Figure 24

*Jules Henri Poincaré, 1854–1912. French scientist.

was standing in the middle of the car, and the events (x and y) happened at each end of his car at a certain instant, which he claimed was the same time for each event, because, standing in the middle of the car, he saw the light from both of these things at the same time. But the man in car A, who happened to be moving with a constant velocity relative to B, saw the same two events, not at the same time, but in fact he saw x first, because the light reached him before the light from y, because he was moving forward. You see that one of the consequences of the principle of symmetry for uniform velocity in a straight line – that word symmetry means that you cannot tell who's view is correct – is that when I talk about everything that is happening in the world 'now', that does not mean anything. If you are moving along at a uniform velocity in a straight line, then the things that happen that appear to you as simultaneous are not the same events as appear simultaneous to me, even though we are passing each other on the instant when I consider the simultaneous event to have happened. We cannot agree what 'now' means at a distance. This means a profound transformation of our ideas of space and time, in order to maintain this principle that uniform velocity in a straight line cannot be detected. Actually what is happening here is that two things which appear from one point of view to be simultaneous, seem from another point of view to be not at the same time, provided they are not at the same place, but are far apart in distance.

You can see that this is very much like the x and y business in space. If I stand facing an audience, then the two sides of the stage on which I stand are on a level with me. They have the same x, but different y. But if I turn round through 90°, and look at the same pair of walls, but from a different point of view, then one is in front of me and one is behind, they have different x′. So it is that the two events which from one point of view seem to be at the same time (same t), from another point of view can seem to be at different times (different t′). A generalization of the two-dimensional rotation that I spoke about was therefore made into space and time, so that time was added to space to make a four-dimensional world. It is not merely an artificial addition, like the explanation given in most of the popular books, which say 'We add time to space, because you cannot only locate a point, you also have to say when'. That is true, but that would not make it real four-dimensional space-time; that just puts the two things together. Real space has, in a sense, the characteristic that its existence is independent of the particular point of view, and that looked at from different points of view a certain amount of 'forward-backward' can get mixed up with 'left-right'. In an analogous way a certain amount of time 'future-past' can get mixed up with a certain amount of space. Space and time must be completely interlocked; after this discovery Minkowski said that 'Space of itself and time of itself shall sink into mere shadows, and only a kind of union between them shall survive'.

I bring this particular example up in such detail because it is really the beginning of the study of symmetries in physical laws. It was Poincaré's suggestion to make this analysis of what you can do to the equations and leave them alone. It was Poincaré's attitude to pay attention to the symmetries of physical laws. The symmetries of translation in space, delay in time, and so on, were not very deep; but the symmetry of uniform velocity in a straight line is very interesting, and has all kinds of consequences. Furthermore, these consequences are extendable into laws that we do not know. For example, by guessing that this principle is true for the disintegration of a mu meson, we can state that we cannot use mu mesons to tell how fast we are going in a space ship either; and thus we know something at least about mu meson disintegration, even though we do not know why the mu meson disintegrates in the first place.

There are many other symmetries, some of them of a very different kind. I will just mention a few. One is that you can replace one atom by another of the same kind and it makes no difference to any phenomenon. Now you may ask 'What do you mean by the same kind?' I can only answer that I mean one which, when replaced by the other one, does not make any difference! It looks as if physicists are always talking nonsense in a way, doesn't it? There are many different kinds of atoms, and if you replace one by one of a different kind it makes a difference, but if you replace one by the same kind it makes no difference, which looks like a circular definition. But the real meaning of the thing is that there *are* atoms of the same kind; that it *is* possible to find groups, classes of atoms, so that you can replace one by another of the same kind and it makes no difference. Since the number of atoms in any tiny little piece of material is 1 followed by 23 noughts or so, it is very important that they are the same, that they are not all different. It is really very interesting that we can classify them into a limited number of a few hundred different types of atom, so the statement that we can replace one atom by another of the same kind has a great amount of content. It has the greatest amount of content in quantum mechanics, but it is impossible for me to explain this here, partly, but only partly, because this lecture is addressed to an audience that is mathematically untrained; it is quite subtle anyway. In quantum mechanics the proposition that you can replace one atom by another of the same kind has marvellous consequences. It produces peculiar phenomena in liquid helium, the liquid that flows through pipes without any resistance, just coasts on for ever. In fact it is the origin of the whole periodic table of the elements, and of the force that keeps me from going through the floor. I cannot go into all this in detail, but I want to emphasize the importance of looking at these principles.

By this time you are probably convinced that all the laws of physics are symmetrical under any kind of change whatsoever, so now I will give a few that do not work. The first one is change of scale. It is not true that if you build an apparatus, and then build another one, with every part made exactly the same, of the same kind of stuff, but twice as big, that it will work in exactly the same way. You who are familiar with atoms are aware of this fact, because if I made the apparatus ten billion times smaller I would only have five atoms in it, and I cannot make, for instance, a machine tool out of only five atoms. It is perfectly obvious if we go that far that we cannot change the scale, but even before the complete awareness of the atomic picture was developed it became apparent that this law is not right. You have probably seen in the newspapers from time to time that somebody has made a cathedral with matchsticks – several floors, and everything more Gothic than any Gothic cathedral has ever been, and more delicate. Why do we never build big cathedrals like that, with great logs, with the same degree of 'ginger cake', the same enormous degree of detail? The answer is that if we did build one it would be so high and so heavy that it would collapse. Ah! But you forgot that when you are comparing two things you must change everything that is in the system. The little cathedral made with matchsticks is attracted to the earth, so to make a comparison the big cathedral should be attracted to an even bigger earth. Too bad. A bigger earth would attract it even more, and the sticks would break even more surely!

This fact that the laws of physics were not unchanged under change of scale was first discovered by Galileo. In discussing the strength of rods and bones, he argued that if you need a bone for a bigger animal – say an animal twice as high, wide, and thick – you will have eight times the weight, so you need a bone that can hold the strength eight times. But what a bone can hold depends on its cross-section, and if you made the bone twice as big it would only have four times the cross-section and would only be able to support four times the weight. In his book *Dialogue on Two New Sciences*, you will see pictures of imaginary bones of enormous dogs, way out of proportion. I suppose Galileo felt that the discovery of the fact that the laws of nature are not unchanged under change of scale was as important as his laws of motion, because they are both put together in the tome on *Two New Sciences*.

Another example of something that is not a symmetry law is the fact that if you are spinning at a uniform angular speed in a space ship, it is not true to say that you cannot tell if you are going around. You can. I might say that you would get dizzy. There are other effects; things get thrown to the walls from the centrifugal force (or however you wish to describe it – I hope there are no teachers of freshman physics in the audience to correct me!). It is possible to tell that the earth is rotating by a pendulum or by a gyroscope, and you are probably aware that various observatories and museums have so-called Foucault* pendulums that prove that the earth is rotating, without looking at the stars. It is possible to tell that we are going around at a uniform angular velocity on the earth without looking outside, because the laws of physics are not unchanged by such a motion.

Many people have proposed that really the earth is rotating relative to the galaxies, and that if we were to turn the galaxies too it would not make any difference. Well, I do not know what would happen if you were to turn the whole universe, and we have at the moment no way to tell. Nor, at the moment, do we have any theory which describes the influence of a galaxy on things here so that it comes out of this theory – in a straightforward way, and not by cheating or forcing – that the inertia for rotation, the effect of rotation, the fact that a spinning bucket of water has a concave surface, is the result of a force from the objects around. It is not known whether this is true. That it should be the case is known as Mach's principle, but that it is the case has not yet been demonstrated. The more direct experimental question is whether, if we are rotating at a uniform velocity relative to the nebulae, we see any effect. The answer is yes. If we are moving in a space ship at a uniform velocity in a straight line relative to the nebulae, do we see any effect? The answer is no. Two different things. We cannot say that all motion is relative. That is not the content of relativity. Relativity says that uniform velocity in a straight line relative to the nebulae is undetectable.

*Jean Bernard Léon Foucault, 1819–68. French physicist.

The next symmetry law that I would like to discuss is an interesting one and has an interesting history. That is the question of reflection in space. I build a piece of apparatus, let us say a clock, and then a short distance away I build another clock, a mirror image of the first. They match each other like two gloves, right and left; each spring which is wound one way in one clock is wound in the opposite way in the other, and so on. I wind up the two clocks, set them in corresponding positions, and then let them tick. The question is, will they always agree with each other? Will all the machinery of one clock go in the mirror image of the other? I do not know what you would guess about that. You would probably guess it is true; most people did. Of course we are not talking about geography. We can distinguish right and left by geography. We can say that if we stand in Florida and look at New York the ocean is on the right. That distinguishes right and left, and if the clock involved the water of the sea then it would not work if we built it the other way because its ticker would not get in the water. In that case what we would have to imagine is that the geography of the earth was turned round too on the other clock; anything that is involved must be turned round. Nor are we interested in history. If you pick up a screw in a machine shop, the chances are it has a right-hand thread; you might argue that the other clock would not be the same because it would be harder to get the screws. But that is just a question of what kind of things we make. Altogether the first guess is likely to be that nothing makes any difference. It turns out that the laws of gravitation are such that it would not make any difference if the clock worked by gravity. The laws of electricity and magnetism are such that if in addition it had electric and magnetic guts, currents and wires and what-not, the corresponding clock would still work. If the clock involved ordinary nuclear reactions to make it run, it would not make any difference either. But there is something that can make a difference, and I will come to it in a moment.

You may know that it is possible to measure the concentration of sugar in water by putting polarized light through the water. If you put a piece of polaroid that lets light through at a certain axis in the water, you find that when you watch the light as it goes through deeper and deeper sugar water you have to turn another piece of polaroid at the other end of the water more and more to the right to let the light through. If you put the light through the solution in the other direction it is still to the right. Here, then, is a difference between right and left. We could use sugar-water and light in the clocks. Suppose we have a tank of water and make light go through and turn our second piece of polaroid so that the light just gets through; then suppose we make the corresponding arrangement in our second clock, hoping the light will turn to the left. It will not; it will still turn to the right and will not get through. By using sugar water our two clocks can be made different!

This is a most remarkable fact, and it seems at first sight to prove that the physical laws are not symmetric for reflection. However, the sugar that we used that time may have been from sugar beet; but sugar is a fairly simple molecule, and it is possible to make it in the laboratory out of carbon dioxide and water, going through lots of stages in between. If you try artificial sugar, which chemically seems to be the same in every way, it does not turn the light. Bacteria eat sugar; if you put bacteria in the artificial sugar water it turns out that they only eat half the sugar, and when the bacteria are finished and you pass polarized light through the remaining sugar water you find it turns to the *left*. The explanation of this is as follows. Sugar is a complicated molecule, a set of atoms in a complicated arrangement. If you make exactly the same arrangement, but with left as right, then every distance between every pair of atoms is the same in one as in the other, the energy of the molecules is exactly the same, and for all chemical phenomena not involving life they are the same. But living creatures find a difference. Bacteria eat one kind and not the other. The sugar that comes from sugar beet is all one kind, all right-hand molecules, and so it turns the light one way. The bacteria can only eat that kind of molecule. When we manufacture the sugar from substances which are not themselves asymmetrical, simple gases, we make both kinds in equal numbers. Then if we introduce the bacteria, they will remove the kind they can eat and the other is left. That is why the light goes through the other way. It is possible to separate the two types by looking through magnifying glasses at the crystals, as Pasteur* discovered. We can definitely show that all this makes sense, and we can separate the sugar ourselves without waiting for the bacteria if we wish to. But the interesting thing is that the bacteria *can* do this. Does this mean that the living processes do not obey the same laws? Apparently not. It seems that in the living creatures there are many, many complicated molecules, and they all have a kind of thread to them. Some of the most characteristic molecules in living creatures are proteins. They have a corkscrew property, and they go to the right. As far as we can tell, if we could make the same things chemically, but to the left rather than to the right, they would not function biologically because when they met the other proteins they would not fit in the same way. A left-hand thread will fit a left-hand thread, but left and right do not fit. The bacteria having a right-hand thread in their chemical insides can distinguish the right and left sugar.

How did they get that way? Physics and chemistry cannot distinguish the molecules, and can only make both kinds. But biology can. It is easy to believe that the explanation is that long ago, when the life processes first began, some accidental molecule got started and propagated itself by reproducing itself, and so on, until after many many years these funny looking blobs, with knobs sticking out with prongs on, stand and yak at each other . . . But we are nothing but the offspring of the first few molecules, and it was an accident of the first few molecules that they happened to form one way instead of the other. It had to be either one or the other, either left or right, and then it reproduced itself,

*Louis Pasteur, 1822–95. French bacteriologist.

and still propagates on and on. It is much like the screws in the machine shop. You use right-hand thread screws to make new right-hand thread screws, and so on. This fact, that all the molecules in living things have exactly the same kind of thread, is probably one of the deepest demonstrations of the uniformity of the ancestry of life, right back to the completely molecular level.

In order to test better this question about whether the laws of physics are the same, right and left, we can put the problem to ourselves this way. Suppose that we were in telephone conversation with a Martian, or an Arcturian, and we wished to describe things on earth to him. First of all, how is he going to understand our words? That question has been studied intensively by Professor Morrison* at Cornell, and he has pointed out that one way would be to start by saying 'tick, one: tick, tick, two: tick, tick, tick, three:' and so on. Pretty soon the guy would catch on to the numbers. Once he understood your number system, you could write a whole sequence of numbers that represent the weights, the proportional weights, of the different atoms in succession, and then say 'hydrogen, 1·008', then deuterium, helium, and so on. After he had sat down with these numbers for a while he would discover that the mathematical ratios were the same as the ratios for the weights of the elements, and that therefore those names must be the names of the elements. Gradually in this way you could build up a common language. Now comes the problem. Suppose, after you get familiar with him, he says, 'You fellows, you're very nice. I'd like to know what you look like'. You start, 'We're about six feet tall', and he says, 'Six feet – how big is a foot?' That is very easy: 'Six feet tall is seventeen thousand million hydrogen atoms high'. That is not a joke – it is a possible way of describing six feet to someone who has no measure – assuming that we cannot send him any samples, nor can we both look at the same objects. If we wish to tell

*Philip Morrison, American professor of physics, 1964, BBC-1, television series 'The Fabric of the Atom'.

him how big we are we can do it. That is because the laws of physics are not unchanged under a scale change, so we can use that fact to determine the scale. We can go on describing ourselves – we are six feet tall, and we are so-and-so bilateral on the outside, and we look like this, and there are these prongs sticking out, etc. Then he says, 'That's very interesting, but what do you look like on the inside?' So we describe the heart and so on, and we say, 'Now put the heart in on the left side'. The question is, how can we tell him which side is the left side? 'Oh', you say, 'We take beet sugar, and put it in water, and it turns . . .' only the trouble is that he has no beets up there. Also we have no way of knowing whether the accidents of evolution on Mars, even if they had produced corresponding proteins to those here, would have started with the oppositely-handed threads. There is no way to tell. After much thought you see that you cannot do it, and so you conclude it is impossible.

About five years ago, however, certain experiments produced all kinds of puzzles. I will not go into detail, but we found ourselves in tighter and tighter difficulties, more and more paradoxical situations, until finally Lee and Yang* proposed that maybe the principle of right and left symmetry – that nature is the same for right and left – is not correct, and that this would help to explain a number of mysteries. Lee and Yang proposed some more direct experiments to demonstrate this, and I will just mention the most direct of all the experiments done.

We take a radioactive disintegration in which, for instance, an electron and a neutrino are emitted – an example, which we have talked about before, is the disintegration of a neutron into a proton, an electron and an anti-neutrino, and there are many radioactivities in which the charge of the nucleus increases by one and an electron comes out. The thing that is interesting is that if you measure the spin – electrons are spinning as they come out – you find out that

*Tsung-Dao Lee and Chen Ning Yang, Chinese physicists, joint Nobel Prize 1957.

they are spinning to the left (as seen from behind – i.e. if they are going south, they turn in the same direction as does the earth). It has a definite significance, that the electron when it comes out of the disintegration is always turning one way, it has a left-hand thread. It is as though in the beta-decay the gun that was shooting out the electron were a rifled gun. There are two ways to rifle a gun; there is the direction 'out', and you have the choice whether you turn it left or right as you go out. The experiment shows that the electron comes from a rifled gun, rifled to the left. Using this fact, therefore, we could ring up our Martian and say, 'Listen, take a radioactive stuff, a neutron, and look at the electron which comes from such a beta-decay. If the electron is going up as it comes out, the direction of its spin is into the body from the back on the *left* side. That defines left. That is where the heart goes'. Therefore it is possible to tell right from left, and thus the law that the world is symmetrical for left and right has collapsed.

The next thing I would like to talk about is the relationship of conservation laws to symmetry laws. In the last lecture we talked about conservation principles, conservation of energy, momentum, angular momentum, and so on. It is extremely interesting that there seems to be a deep connection between the conservation laws and the symmetry laws. This connection has its proper interpretation, at least as we understand it today, only in the knowledge of quantum mechanics. Nevertheless I will show you one demonstration of this.

If we assume that the laws of physics are describable by a minimum principle, then we can show that if a law is such that you can move all the equipment to one side, in other words if it is translatable in space, then there must be conservation of momentum. There is a deep connection between the symmetry principles and the conservation laws, but that connection requires that the minimum principle be assumed. In the second lecture we discussed one way of describing physical laws by saying that a particle goes from one place to another in a given length of time by trying different paths. There is a certain quantity which, perhaps misleadingly, happens to be called the action. When you calculate the action on the various paths you will find that for the actual path taken this quantity is always smaller than for any other. That way of describing the laws of nature is to say that the action of certain mathematical formulae is least for the actual path of all the possible paths. Another way of saying a thing is least is to say that if you move the path a little bit at first it does not make any difference. Suppose you were walking around on hills – but smooth hills, since the mathematical things involved correspond to smooth things – and you come to a place where you are lowest, then I say that if you take a small step forward you will not change your height. When you are at the lowest or at the highest point, a step does not make any difference in the altitude in first approximation, whereas if you are on a slope you can walk down the slope with a step and then if you take the step in the opposite direction you walk up. That is the key to the reason why, when you are at the lowest place, taking a step does not make much difference, because if it did make any difference then if you took a step in the opposite direction you would go down. Since this is the lowest point and you cannot go down, your first approximation is that the step does not make any difference. We therefore know that if we move a path a little bit it does not make any difference to the action on a first approximation. We draw a path, A to B (fig. 25), and now I want you to consider the following possible other path. First we jump immediately over to another place near by, C, then we move on exactly the corresponding path to another point, which we will call D, which is displaced the same amount, of course, because it is the corresponding path. Now we have just discovered that the laws of nature are such that the total amount of action going on the ACDB path is the same in the first approximation to that original path AB – that is from the minimum principle, when it is the real motion. I will tell you something else. The action on the original path, A to B, is the same as the action from C to D if the

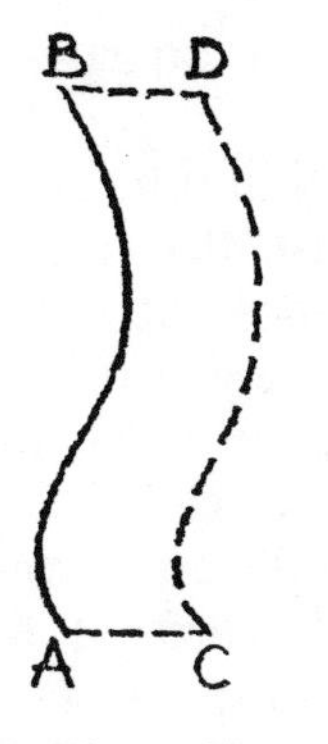

Figure 25

world is the same when you move everything over, because the difference of these two is only that you have moved everything over. So if the symmetry principle of translation in space is right, then the action on the direct path between A and B is the same as that on the direct path between C and D. However for the true motion the total action on the indirect path ACDB is very nearly the same as on the direct path AB, and therefore the same as just the part C to D. This indirect action is the sum of three parts – the action going A to C, that of C to D, plus that from D to B. So, subtracting equals from equals, you can probably see that the contribution from A to C and that from D to B must add up to zero. But in the motion for one of these sections we are going one way, and for the other the opposite way. If we take the contribution of A to C, thinking of it as an effect of moving one way, and the contribution of D to B as B to D, taking the opposite sign because it is the other way, we see that there is a quantity A to C which has to match the quantity B to D to cancel off. This is the effect on the action of a tiny step in the B to D direction. That quantity, the effect on the action of a small step to the right, is the same at the beginning (A to C) as at the end (B to D). There is a quantity, therefore, that does not change as time goes on, provided the minimum principle works, and the symmetry principle of displacement in space is right. This quantity which does not change (the effect on the action of a small step to one side) is in fact exactly the momentum that we discussed in the last lecture. This shows the relation of symmetry laws to conservation laws, assuming the laws obey a principle of least action. They satisfy a principle of least action, it turns out, because they come from quantum mechanics. That is why I said that in the last analysis the connection of symmetry laws to conservation laws comes from quantum mechanics.

The corresponding argument for delay in time comes out as the conservation of energy. The case that rotation in space does not make any difference comes out as the conservation of angular momentum. That we can reflect without any change in effect does not come out to be anything simple in the classical sense. People have called it parity, and they have a conservation law called the conservation of parity, but these are just complicated words. I have to mention the conservation of parity, because you may have read in the papers that the law of the conservation of parity has been proved wrong. It would have been much easier to understand if what had been written was that the principle that you cannot distinguish right from left has been proved wrong.

Whilst I am talking about symmetries, one thing I would like to tell you is that there are a few new problems. For instance, for every particle there is an anti-particle: for an electron this is a positron, for a proton an anti-proton. We can in principle make what we call anti-matter, in which every atom has its corresponding anti-pieces put together. The hydrogen atom is a proton and an electron; if we take an anti-proton, which is electrically negative, and a positron, and put them together, they also will make a kind of hydrogen atom, an anti-hydrogen atom. Anti-hydrogen atoms have never in fact been made, but it has been figured out that in principle it would work, and that we could make all kinds of anti-matter in the same manner. What we would ask now is whether the anti-matter works in the same way as matter, and as far as we know it does. One of the laws of symmetry is that if we made stuff out of anti-matter it would behave in the same way as if we made the corresponding stuff out of matter. Of course if they came together they would annihilate one another and there would be sparks.

It always has been believed that matter and anti-matter have the same laws. However, now we know that the left and right symmetry appears wrong, an important question comes. If I look at the neutron disintegration, but with anti-matter – an anti-neutron goes into an anti-proton plus an anti-electron (also called a positron), plus a neutrino – the question is, does it behave in the same way, in the sense that the positron will come out with a left-hand thread, or does it behave the other way? Until a few months ago we believed that it behaves the opposite way, and that the anti-matter (positron) goes to the right where matter (electron) goes to the left. In that case we cannot really tell the Martian which is right and left, because if he happens to be made out of anti-matter, when he does his experiment his electrons will be positrons, and they will come up spinning the wrong way and he will put the heart on the wrong side. Suppose you telephone the Martian, and you explain how to make a man; he makes one, and it works. Then you explain to him also all our social conventions. Finally, after he tells us how to build a sufficiently good space ship, you go to meet this man, and you walk up to him and put out your right hand to shake hands. If he puts out his right hand, O.K., but if he puts out his left hand watch out . . . the two of you will annihilate with each other!

I wish I could tell you about a few more symmetries, but they become more difficult to explain. There are also some very remarkable things, which are the near-symmetries. For instance, the remarkable feature of the fact that we can distinguish right and left is that we can only do so with a very weak effect, with this beta-disintegration. What this means is that nature is 99·99 per cent indistinguishable right from left, but that there is just one little piece, one little characteristic phenomenon, which is completely different, in the sense that it is absolutely lop-sided. This is a mystery that no one has the slightest idea about yet.

5

The Distinction of Past and Future

It is obvious to everybody that the phenomena of the world are evidently irreversible. I mean things happen that do not happen the other way. You drop a cup and it breaks, and you can sit there a long time waiting for the pieces to come together and jump back into your hand. If you watch the waves of the sea breaking, you can stand there and wait for the great moment when the foam collects together, rises up out of the sea, and falls back farther out from the shore – it would be very pretty!

The demonstration of this in lectures is usually made by having a section of moving picture in which you take a number of phenomena, and run the film backwards, and then wait for all the laughter. The laughter just means this would not happen in the real world. But actually that is a rather weak way to put something which is as obvious and deep as the difference between the past and the future; because even without an experiment our very experiences inside are completely different for past and future. We remember the past, we do not remember the future. We have a different kind of awareness about what might happen than we have of what probably has happened. The past and the future look completely different psychologically, with concepts like memory and apparent freedom of will, in the sense that we feel that we can do something to affect the future, but none of us, or very few of us, believe that there is anything we can do to affect the past. Remorse and regret and hope and so forth are all words which distinguish perfectly obviously the past and the future.

Now if the world of nature is made of atoms, and we too are made of atoms and obey physical laws, the most obvious interpretation of this evident distinction between past and future, and this irreversibility of all phenomena, would be that some laws, some of the motion laws of the atoms, are going one way – that the atom laws are not such that they can go either way. There should be somewhere in the works some kind of a principle that uxles only make wuxles, and never vice versa, and so the world is turning from uxley character to wuxley character all the time – and this one-way business of the interactions of things should be the thing that makes the whole phenomena of the world seem to go one way.

But we have not found this yet. That is, in all the laws of physics that we have found so far there does not seem to be any distinction between the past and the future. The moving picture should work the same going both ways, and the physicist who looks at it should not laugh.

Let us take the law of gravitation as our standard example. If I have a sun and a planet, and I start the planet off in some direction, going around the sun, and then I take a moving picture, and run the moving picture backwards and look at it, what happens? The planet goes around the sun, the opposite way of course, keeps on going around in an ellipse. The speed of the planet is such that the area swept out by the radius is always the same in equal times. In fact it just goes exactly the way it ought to go. It cannot be distinguished from going the other way. So the law of gravitation is of such a kind that the direction does not make any difference; if you show any phenomenon involving only gravitation running backwards on a film it will look perfectly satisfactory. You can put it more precisely this way. If all the particles in a more complicated system were to have every one of their speeds reversed suddenly, then the thing would just unwind through all the things that it wound up into. If you have a lot of particles doing something, and then you suddenly reverse the speed, they will completely undo what they did before.

This is in the law of gravitation, which says that the velocity changes as a result of the forces. If I reverse the time, the forces are not changed, and so the changes in velocity are not altered at corresponding distances. So each velocity then has a succession of alterations made in exactly the reverse of the way that they were made before, and it is easy to prove that the law of gravitation is time-reversible.

The law of electricity and magnetism? Time reversible. The laws of nuclear interaction? Time reversible as far as we can tell. The laws of beta-decay that we talked about at a previous time? Also time reversible? The difficulty of the experiments of a few months ago, which indicate that there is something the matter, some unknown about the laws, suggests the possibility that in fact beta-decay may not also be time reversible, and we shall have to wait for more experiments to see. But at least the following is true. Beta-decay (which may or may not be time reversible) is a very unimportant phenomenon for most ordinary circumstances. The possibility of my talking to you does not depend upon beta-decay, although it does depend on chemical interactions, it depends on electrical forces, not much on nuclear forces at the moment, but it depends also on gravitation. But I am one-sided – I speak, and a voice goes out into the air, and it does not come sucking back into my mouth when I open it – and this irreversibility cannot be hung on the phenomenon of beta-decay. In other words, we believe that most of the ordinary phenomena in the world, which are produced by atomic motions, are according to laws which can be completely reversed. So we will have to look some more to find the explanation of the irreversibility.

If we look at our planets moving around the sun more carefully, we soon find that all is not quite right. For example, the Earth's rotation on its axis is slightly slowing down. It is due to tidal friction, and you can see that friction is something which is obviously irreversible. If I take a heavy weight on the floor, and push it, it will slide and stop. If I stand and wait, it does not suddenly start up and speed up and come into my hand. So the frictional effect seems to be irreversible. But a frictional effect, as we discussed at another time, is the result of the enormous complexity of the interactions of the weight with the wood, the jiggling of the atoms inside. The organized motion of the weight is changed into disorganized, irregular wiggle-waggles of the atoms in the wood. So therefore we should look at the thing more closely.

As a matter of fact, we have here the clue to the apparent irreversibility. I will take a simple example. Suppose we have blue water, with ink, and white water, that is without ink, in a tank, with a little separation, and then we pull out the separation very delicately. The water starts separate, blue on one side and white on the other side. Wait a while. Gradually the blue mixes up with the white, and after a while the water is 'luke blue', I mean it is sort of fifty-fifty, the colour uniformly distributed throughout. Now if we wait and watch this for a long time, it does not by itself separate. (You could *do* something to get the blue separated again. You could evaporate the water and condense it somewhere else, and collect the blue dye and dissolve it in half the water, and put the thing back. But while you were doing all that you yourself would be causing irreversible phenomena somewhere else.) By itself it does not go the other way.

That gives us some clue. Let us look at the molecules. Suppose that we take a moving picture of the blue and white water mixing. It will look funny if we run it backwards, because we shall start with uniform water and gradually the thing will separate – it will be obviously nutty. Now we magnify the picture, so that every physicist can watch, atom by atom, to find out what happens irreversibly – where the laws of balance of forward and backward break down. So you start, and you look at the picture. You have atoms of two different kinds (it's ridiculous, but let's call them blue and white) jiggling all the time in thermal motion. If we were to start at the beginning we should have mostly atoms of one kind on one side, and atoms of the other kind on the other side. Now these atoms are jiggling around, billions and billions of them, and if we start them with one kind all on one side, and the other kind on the other side, we see that in their perpetual irregular motions they will get mixed up, and that is why the water becomes more or less uniformly blue.

Let us watch any one collision selected from that picture, and in the moving picture the atoms come together this way and bounce off that way. Now run that section of the film backwards, and you find the pair of molecules moving together the other way and bouncing off this way. And the physicist looks with his keen eye, and measures everything, and says, 'That's all right, that's according to the laws of physics. If two molecules came this way they would bounce this way'. It is reversible. The laws of molecular collision are reversible.

So if you watch too carefully you cannot understand it at all, because every one of the collisions is absolutely reversible, and yet the whole moving picture shows something absurd, which is that in the reversed picture the molecules start in the mixed condition – blue, white, blue, white, blue, white – and as time goes on, through all the collisions, the blue separates from the white. But they cannot do that – it is not natural that the accidents of life should be such that the blues will separate themselves from the whites. And yet if you watch this reversed movie very carefully every collision is O.K.

Well you see that all there is to it is that the irreversibility is caused by the general accidents of life. If you start with a thing that is separated and make irregular changes, it does get more uniform. But if it starts uniform and you make irregular changes, it does not get separated. It *could* get separated. It is not against the laws of physics that the molecules bounce around so that they separate. It is just unlikely. It would never happen in a million years. And that is the answer. Things are irreversible only in a sense that going one way is likely, but going the other way, although it is possible and is according to the laws of physics, would not happen in a million years. It is just ridiculous to expect that if you sit there long enough the jiggling of the atoms will separate a uniform mixture of ink and water into ink on one side and water on the other.

Now if I had put a box around my experiment, so that there were only four or five molecules of each kind in the box, as time went on they would get mixed up. But I think you could believe that, if you kept watching, in the perpetual irregular collisions of these molecules, after some time – not necessarily a million years, maybe only a year – you would see that, accidentally they would get back more or less to their original state, at least in the sense that if I put a barrier through the middle, all the whites would be on one side and all the blues on the other. It is not impossible. However, the actual objects with which we work have not only four or five blues and whites. They have four or five million, million, million, million, which are all going to get separated like this. And so the apparent irreversibility of nature does not come from the irreversibility of the fundamental physical laws; it comes from the characteristic that if you start with an ordered system, and have the irregularities of nature, the bouncing of molecules, then the thing goes one way.

Therefore the next question is – how did they get ordered in the first place? That is to say, why is it possible to start with the ordered? The difficulty is that we start with an ordered thing, and we do not end with an ordered thing. One of the rules of the world is that the thing goes from an ordered condition to a disordered. Incidentally, this word order, like the word disorder, is another of these terms of physics which are not exactly the same as in ordinary life. The order need not be interesting to you as human beings, it is just that there is a definite situation, all on one side and all on the other, or they are mixed up – and that is ordered and disordered.

The question, then, is how the thing gets ordered in the first place, and why, when we look at any ordinary situation, which is only partly ordered, we can conclude that it probably came from one which was more ordered. If I look at a tank of water, in which the water is very dark blue on one side and very clear white on the other, and a sort of bluish colour in between, and I know that the thing has been left alone for twenty or thirty minutes, then I will guess that it got this way because the separation was more complete in the past. If I wait longer, then the blue and white will get more intermixed, and if I know that this thing has been left alone for a sufficiently long time, I can conclude something about the past condition. The fact that it is 'smooth' at the sides can only arise because it was much more satisfactorily separated in the past; because if it were not more satisfactorily separated in the past, in the time since then it would have become more mixed up than it is. It is therefore possible to tell, from the present, something about the past.

In fact physicists do not usually do this much. Physicists like to think that all you have to do is say, 'These are the conditions, now what happens next?' But all our sister sciences have a completely different problem: in fact all the other things that are studied – history, geology, astronomical history – have a problem of this other kind. I find they are able to make predictions of a completely different type from those of a physicist. A physicist says, 'In this condition I'll tell you what will happen next'. But a geologist will say something like this – 'I have dug in the ground and I have found certain kinds of bones. I predict that if you dig in the ground you will find a similar kind of bones'. The historian, although he talks about the past, can do it by talking about the future. When he says that the French Revolution was in 1789, he means that if you look in another book about the French Revolution you will find the same date. What he does is to make a kind of prediction about something that he has never looked at before, documents that have still to be found. He predicts that the documents in which there is something written about Napoleon will coincide with what is written in the other documents. The question is how that is possible – and the only way that is possible is to suggest that the past of the world was more organized in this sense than the present.

Some people have proposed that the way the world became ordered is this. In the beginning the whole universe was just irregular motions, like the mixed water. We saw that if you waited long enough, with very few atoms, the water could have got separated accidentally. Some physicists (a century ago) suggested that all that has happened is that the world, this system that has been going on and going on, fluctuated. (That is the term used when it gets a little out of the ordinary uniform condition.) It fluctuated, and now we are watching the fluctuation undo itself again. You may say, 'But look how long you would have to wait for such a fluctuation.' I know, but if it did not fluctuate far enough to be able to produce evolution, to be able to produce an intelligent person, we would not have noticed it. So we had to keep waiting until we were alive to notice it – we had to have at least that big a fluctuation. But I believe this theory to be incorrect. I think it is a ridiculous theory for the following reason. If the world were much bigger, and the atoms were all over the place starting from a completely mixed up condition, then if I happened to look only at the atoms in one place, and I found the atoms there separated, I would have no way to conclude that the atoms anywhere else would be separated. In fact if the thing were a fluctuation, and I noticed something odd, the most likely way that it got there would be that there was nothing odd anywhere else. That is, I would have to borrow odds, so to speak, to get the thing lopsided, and there is no use borrowing too much. In the experiment with the blue and white water, when eventually the few molecules in the box became separated, the most likely condition of the rest of the water would still be mixed up. And therefore, although when we look at the stars and we look at the world we see everything is ordered, if there were a fluctuation, the prediction would be that if we looked at a place where we have not looked before, it would be disordered and a mess. Although the separation of the matter into stars which are hot and space which is cold, which we have seen, could be a fluctuation, then in places where we have not looked we would expect to find that the stars are not separated from space. And since we always make the prediction that in a place where we have not looked we shall see stars in a similar condition, or find the same statement about Napoleon, or that we shall see bones like the bones that we have seen before, the success of all those sciences indicates that the world did not come from a fluctuation, but came from a condition which was more separated, more organized, in the past than at the present time. Therefore I think it necessary to add to the physical laws the hypothesis that in the past the universe was more ordered, in the technical sense, than it is today – I think this is the additional statement that is needed to make sense, and to make an understanding of the irreversibility.

That statement itself is of course lopsided in time; it says that something about the past is different from the future. But it comes outside the province of what we ordinarily call physical laws, because we try today to distinguish between the statement of the physical laws which govern the rules by which the universe develops, and the law which states the condition that the world was in in the past. This is considered to be astronomical history – perhaps some day it will also be a part of physical law.

Now there are a number of interesting features of irreversibility which I would like to illustrate. One of them is to see how, exactly, an irreversible machine really works.

Suppose that we build something that we know ought to work only one way – and what I am going to build is a wheel with a ratchet on it – a saw-toothed wheel, with sharp up notches, and relatively slow down notches, all the way round. The wheel is on a shaft, and then there is a little

Figure 26

pawl, which is on a pivot and which is held down by a spring (fig. 26).

Now the wheel can only turn one way. If you try to turn it the other way, the straight-edged parts of the teeth get jammed against the pawl and it does not go, whereas if you turn it the other way it just goes right over the teeth, snap, snap, snap. (You know the sort of thing: they use them in clocks, and a watch has this kind of thing inside so that you can only wind it one way, and after you have wound it, it holds the spring.) It is completely irreversible in the sense that the wheel can only turn one way.

Now it has been imagined that this irreversible machine, this wheel that can only turn one way, could be used for a very useful and interesting thing. As you know, there is a perpetual irregular motion of molecules, and if you build a very delicate instrument it will always jiggle because it is being bombarded irregularly by the air molecules in the neighbourhood. Well that is very clever, so we will connect the wheel with a shaft that has four vanes, like this (fig. 27).

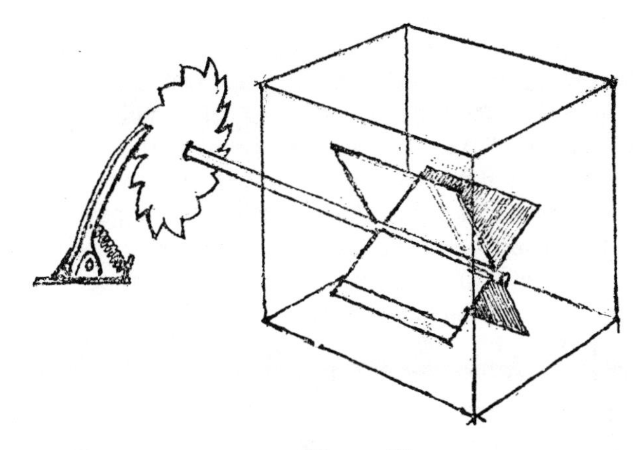

Figure 27

They are in a box of gas, and they are bombarded all the time by the molecules irregularly, so the vanes are pushed sometimes one way, sometimes the other way. But when the vanes are pushed one way the thing gets jammed by the ratchet, and when the vanes are pushed the other way, it goes around, and so we find the wheel perpetually going around, and we have a kind of perpetual motion. That is because the ratchet wheel is irreversible.

But actually we have to look into things in more detail. The way this works is that when the wheel goes one way it lifts the pawl up and then the pawl snaps down against the tooth. Then it will bounce off, and if it is perfectly elastic it will go bounce, bounce, bounce, all the time, and the wheel can just go down and around the other way when the pawl accidentally bounces up. So this will not work unless it is true that when the pawl comes down it sticks, or stops, or bounces and cuts out. If it bounces and cuts out there must be what we call damping, or friction, and in the falling down and bouncing and stopping, which is the only way this will work one-way, heat is generated by the friction, so the wheel will get hotter and hotter. However, when it begins to get quite warm something else happens. Just as there is Brownian motion, or irregular motions, in the gas round the vanes, so whatever this wheel and pawl are made of, the parts that they are made of, are getting hotter, and are beginning to move in a more irregular fashion. The time comes when the wheel is so hot that the pawl is simply jiggling because of the molecular motions of the things inside it, and so it bounces up and down on the wheel because of molecular motion, the same thing as was making the vane turn round. In bouncing up and down on the wheel it is up as much as it is down, and the tooth can go either way. We no longer have a one-way device. As a matter of fact, the thing can be driven backwards! If the wheel is hot and the vane part is cold, the wheel that you thought would go only one way will go the other way, because every time the pawl comes down it comes down on an inclined plane on the toothed wheel, and so pushes the wheel 'backwards'. Then it bounces up again, comes down on another inclined plane, and so on. So if the wheel is hotter than the vanes it will go the wrong way.

What has this got to do with the temperature of the gas round the vanes? Suppose we did not have that part at all. Then if the wheel is pushed forward by the pawl falling on an inclined plane, the next thing that will happen is that the straight vertical side of the tooth will bounce against the pawl and the wheel will bounce back. In order to prevent the wheel from bouncing back we put a damper on it and put vanes in the air, so it will be slowed down and not bounce freely. Then it will go only one way, but the wrong way, and so it turns out that no matter how you design it, a wheel like this will go one way if one side is hotter and the other way if the other side is hotter. But after there is a heat exchange between the two, and everything is calmed down, so that the vane and the wheel have come to be at the same temperature, it will neither go the one way nor the other on the average. That is the technical way in which the phenomena of nature will go one way as long as they are out of equilibrium, as long as one side is quieter than the other, or one side is bluer than the other.

The conservation of energy would let us think that we have as much energy as we want. Nature never loses or gains energy. Yet the energy of the sea, for example, the thermal motion of all the atoms in the sea, is practically unavailable to us. In order to get that energy organized, herded, to make it available for use, we have to have a difference in temperature, or else we shall find that although the energy is there we cannot make use of it. There is a great difference between energy and availability of energy. The energy of the sea is a large amount, but it is not available to us.

The conservation of energy means that the total energy in the world is kept the same. But in the irregular jigglings that energy can be spread about so uniformly that, in certain circumstances, there is no way to make more go one way than the other – there is no way to control it any more.

I think that by an analogy I can give some idea of the difficulty, in this way. I do not know if you have ever had the experience – I have – of sitting on the beach with several towels, and suddenly a tremendous downpour comes. You pick up the towels as quickly as you can, and run into the bathhouse. Then you start to dry yourself, and you find that this towel is a little wet, but it is drier than you are. You keep drying with this one until you find it is too wet – it is wetting you as much as drying you – and you try another one; and pretty soon you discover a horrible thing – that all the towels are damp and so are you. There is no way to get any drier, even though you have many towels, because there is no difference in some sense between the wetness of the towels and the wetness of yourself. I could invent a kind of quantity which I could call 'ease of removing water'. The towel has the same ease of removing water from it as you have, so when you touch yourself with the towel, as much water comes off the towel on to you as comes from you to the towel. It does not mean there is the same amount of water in the towel as there is on you – a big towel will have more water in it than a little towel – but they have the same dampness. When things get to the same dampness then there is nothing you can do any longer.

Now the water is like the energy, because the total amount of water is not changing. (If the bathhouse door is open and you can run into the sun and get dried out, or find another towel, then you're saved, but suppose everything is closed, and you can't get away from these towels or get any new towels.) In the same way if you imagine a part of the world that is closed, and wait long enough, in the accidents of the world the energy, like the water, will be distributed over all of the parts evenly until there is nothing left of one-way-ness, nothing left of the real interest of the world as we experience it.

Thus in the ratchet and pawl and vanes situation, which is a limited one, in which nothing else is involved, the temperatures gradually become equal on both sides, and the wheel does not go round either one way or the other. In the same way the situation is that if you leave any system long enough it gets the energy thoroughly mixed up in it, and no more energy is really available to do anything.

Incidentally, the thing that corresponds to the dampness

or the 'ease of removing water' is called the temperature, and although I say when two things are at the same temperature things get balanced, it does not mean they have the same energy in them; it means that it is just as easy to pick energy off one as to pick it off the other. Temperature is like 'ease of removing energy'. So if you sit them next to each other, nothing apparently happens; they pass energy back and forth equally, but the net result is nothing. So when things have become all of the same temperature, there is no more energy available to do anything. The principle of irreversibility is that if things are at different temperatures and are left to themselves, as time goes on they become more and more at the same temperature, and the availability of energy is perpetually decreasing.

This is another name for what is called the entropy law, which says the entropy is always increasing. But never mind the words; stated the other way, the availability of energy is always decreasing. And that is a characteristic of the world, in the sense that it is due to the chaos of molecular irregular motions. Things of different temperature, if left to themselves, tend to become of the same temperature. If you have two things at the same temperature, like water on an ordinary stove without a fire under it, the water is not going to freeze and the stove get hot. But if you have a hot stove with ice, it goes the other way. So the one-way-ness is always to the loss of the availability of energy.

That is all I want to say on the subject, but I want to make a few remarks about some characteristics. Here we have an example in which an obvious effect, the irreversibility, is not an obvious consequence of the laws, but is in fact rather far from the basic laws. It takes a lot of analysis to understand the reason for it. The effect is of first importance in the economy of the world, in the real behaviour of the world in all obvious things. My memory, my characteristics, the difference between past and future, are completely involved in this, and yet the understanding of it is not prima facie available by knowing about the laws. It takes a lot of analysis.

It is often the case that the laws of physics do not have an obvious direct relevance to experience, but that they are abstract from experience to varying degrees. In this particular case, the fact that the laws are reversible although the phenomena are not is an example.

There are often great distances between the detailed laws and the main aspects of real phenomena. For example, if you watch a glacier from a distance, and see the big rocks falling into the sea, and the way the ice moves, and so forth, it is not really essential to remember that it is made out of little hexagonal ice crystals. Yet if understood well enough the motion of the glacier is in fact a consequence of the character of the hexagonal ice crystals. But it takes quite a while to understand all the behaviour of the glacier (in fact nobody knows enough about ice yet, no matter how much they've studied the crystal). However the hope is that if we do understand the ice crystal we shall ultimately understand the glacier.

In fact, although we have been talking in these lectures about the fundaments of the physical laws, I must say immediately that one does not, by knowing all the fundamental laws as we know them today, immediately obtain an understanding of anything much. It takes a while, and even then it is only partial. Nature, as a matter of fact, seems to be so designed that the most important things in the real world appear to be a kind of complicated accidental result of a lot of laws.

To give an example, nuclei, which involve several nuclear particles, protons and neutrons, are very complicated. They have what we call energy levels, they can sit in states or conditions of different energy values, and various nuclei have various energy levels. And it's a complicated mathematical problem, which we can only partly solve, to find the position of the energy levels. The exact position of the levels is obviously a consequence of an enormous complexity and therefore there is no particular mystery about the fact that nitrogen, with 15 particles inside, happens to have a level at 2·4 million volts, and another level at 7·1, and so on. But the remarkable thing about nature is that the whole universe in its character depends upon precisely the position of one particular level in one particular nucleus. In the carbon^{12} nucleus, it so happens, there is a level at 7·82 million volts. And that makes all the difference in the world.

The situation is the following. If we start with hydrogen, and it appears that at the beginning the world was practically all hydrogen, then as the hydrogen comes together under gravity and gets hotter, nuclear reactions can take place, and it can form helium, and then the helium can combine only partially with the hydrogen and produce a few more elements, a little heavier. But these heavier elements disintegrate right away back into helium. Therefore for a while there was a great mystery about where all the other elements in the world came from, because starting with hydrogen the cooking processes inside the stars would not make much more than helium and less than half a dozen other elements. Faced with this problem, Professors Hoyle and Salpeter* said that there is one way out. If three helium atoms could come together to form carbon, we can easily calculate how often that should happen in a star. And it turns out that it should never happen, except for one possible accident – if there happened to be an energy level at 7·82 million volts in carbon, then the three helium atoms would come together and before they came apart, would stay together a little longer on the average than they would do if there were no level at 7·82. And staying there a little longer, there would be enough time for something else to happen, and to make other elements. If there was a level at 7·82 million volts in carbon, then we could understand where all the other elements in the periodic table came from. And so, by a back-handed, upside-down argument, it was predicted that there is in carbon a level at 7·82 million volts; and experiments in the laboratory showed that indeed there is. Therefore the existence in the world of all these other elements is very

*Fred Hoyle, British astronomer, Cambridge. Edwin Salpeter, American physicist, Cornell University.

closely related to the fact that there is this particular level in carbon. But the position of this particular level in carbon seems to us, knowing the physical laws, to be a very complicated accident of 12 complicated particles interacting. This example is an excellent illustration of the fact that an understanding of the physical laws does not necessarily give you an understanding of things of significance in the world in any direct way. The details of real experience are often very far from the fundamental laws.

We have a way of discussing the world, when we talk of it at various hierarchies, or levels. Now I do not mean to be very precise, dividing the world into definite levels, but I will indicate, by describing a set of ideas, what I mean by hierarchies of ideas.

For example, at one end we have the fundamental laws of physics. Then we invent other terms for concepts which are approximate, which have, we believe, their ultimate explanation in terms of the fundamental laws. For instance, 'heat'. Heat is supposed to be jiggling, and the word for a hot thing is just the word for a mass of atoms which are jiggling. But for a while, if we are talking about heat, we sometimes forget about the atoms jiggling – just as when we talk about the glacier we do not always think of the hexagonal ice and the snowflakes which originally fell. Another example of the same thing is a salt crystal. Looked at fundamentally it is a lot of protons, neutrons, and electrons; but we have this concept 'salt crystal', which carries a whole pattern already of fundamental interactions. An idea like pressure is the same.

Now if we go higher up from this, in another level we have properties of substances – like 'refractive index', how light is bent when it goes through something; or 'surface tension', the fact that water tends to pull itself together, both of which are described by numbers. I remind you that we have to go through several laws down to find out that it is the pull of the atoms, and so on. But we still say 'surface tension', and do not always worry, when discussing surface tension, about the inner workings.

On, up in the hierarchy. With the water we have waves, and we have a thing like a storm, the word 'storm' which represents an enormous mass of phenomena, or a 'sun spot', or 'star', which is an accumulation of things. And it is not worth while always to think of it way back. In fact we cannot, because the higher up we go the more steps we have in between, each one of which is a little weak. We have not thought them all through yet.

As we go up in this hierarchy of complexity, we get to things like muscle twitch, or nerve impulse, which is an enormously complicated thing in the physical world, involving an organization of matter in a very elaborate complexity. Then come things like 'frog'.

And then we go on, and we come to words and concepts like 'man', and 'history', or 'political expediency', and so forth, a series of concepts which we use to understand things at an ever higher level.

And going on, we come to things like evil, and beauty, and hope . . .

Which end is nearer to God; if I may use a religious metaphor. Beauty and hope, or the fundamental laws? I think that the right way, of course, is to say that what we have to look at is the whole structural interconnection of the thing; and that all the sciences, and not just the sciences but all the efforts of intellectual kinds, are an endeavour to see the connections of the hierarchies, to connect beauty to history, to connect history to man's psychology, man's psychology to the working of the brain, the brain to the neural impulse, the neural impulse to the chemistry, and so forth, up and down, both ways. And today we cannot, and it is no use making believe that we can, draw carefully a line all the way from one end of this thing to the other, because we have only just begun to see that there is this relative hierarchy.

And I do not think either end is nearer to God. To stand at either end, and to walk off that end of the pier only, hoping that out in that direction is the complete understanding, is a mistake. And to stand with evil and beauty and hope, or to stand with the fundamental laws, hoping that way to get a deep understanding of the whole world, with that aspect alone, is a mistake. It is not sensible for the ones who specialize at one end, and the ones who specialize at the other end, to have such disregard for each other. (They don't actually, but people say they do.) The great mass of workers in between, connecting one step to another, are improving all the time our understanding of the world, both from working at the ends and working in the middle, and in that way we are gradually understanding this tremendous world of interconnecting hierarchies.

6

Probability and Uncertainty – the Quantum Mechanical view of Nature

In the beginning of the history of experimental observation, or any other kind of observation on scientific things, it is intuition, which is really based on simple experience with everyday objects, that suggests reasonable explanations for things. But as we try to widen and make more consistent our description of what we see, as it gets wider and wider and we see a greater range of phenomena, the explanations become what we call laws instead of simple explanations. One odd characteristic is that they often seem to become more and more unreasonable and more and more intuitively far from obvious. To take an example, in the relativity theory the proposition is that if you think two things occur at the same time that is just your opinion, someone else could conclude that of those events one was before the other, and that therefore simultaneity is merely a subjective impression.

There is no reason why we should expect things to be otherwise, because the things of everyday experience involve large numbers of particles, or involve things moving very slowly, or involve other conditions that are special and represent in fact a limited experience with nature. It is a small section only of natural phenomena that one gets from direct experience. It is only through refined measurements and careful experimentation that we can have a wider vision. And then we see unexpected things: we see things that are far from what we would guess – far from what we could have imagined. Our imagination is stretched to the utmost, not, as in fiction, to imagine things which are not really there, but just to comprehend those things which *are* there. It is this kind of situation that I want to discuss.

Let us start with the history of light. At first light was assumed to behave very much like a shower of particles, of corpuscles, like rain, or like bullets from a gun. Then with further research it was clear that this was not right, that the light actually behaved like waves, like water waves for instance. Then in the twentieth century, on further research, it appeared again that light actually behaved in many ways like particles. In the photo-electric effect you could count these particles – they are called photons now. Electrons, when they were first discovered, behaved exactly like particles or bullets, very simply. Further research showed, from electron diffraction experiments for example, that they behaved like waves. As time went on there was a growing confusion about how these things really behaved – waves or particles, particles or waves? Everything looked like both.

This growing confusion was resolved in 1925 or 1926 with the advent of the correct equations for quantum mechanics. Now we know how the electrons and light behave. But what can I call it? If I say they behave like particles I give the wrong impression; also if I say they behave like waves. They behave in their own inimitable way, which technically could be called a quantum mechanical way. They behave in a way that is like nothing that you have ever seen before. Your experience with things that you have seen before is incomplete. The behaviour of things on a very tiny scale is simply different. An atom does not behave like a weight hanging on a spring and oscillating. Nor does it behave like a miniature representation of the solar system with little planets going around in orbits. Nor does it appear to be somewhat like a cloud or fog of some sort surrounding the nucleus. It behaves like nothing you have ever seen before.

There is one simplification at least. Electrons behave in this respect in exactly the same way as photons; they are both screwy, but in exactly the same way.

How they behave, therefore, takes a great deal of imagination to appreciate, because we are going to describe something which is different from anything you know about. In that respect at least this is perhaps the most difficult lecture of the series, in the sense that it is abstract, in the sense that it is not close to experience. I cannot avoid that. Were I to give a series of lectures on the character of physical law, and to leave out from this series the description of the actual behaviour of particles on a small scale, I would certainly not be doing the job. This thing is completely characteristic of all of the particles of nature, and of a universal character, so if you want to hear about the character of physical law it is essential to talk about this particular aspect.

It will be difficult. But the difficulty really is psychological and exists in the perpetual torment that results from your saying to yourself, 'But how can it be like that?' which is a reflection of uncontrolled but utterly vain desire to see it in terms of something familiar. I *will not* describe it in terms of an analogy with something familiar; I will simply describe it. There was a time when the newspapers said that only twelve men understood the theory of relativity. I do not believe there ever was such a time. There might have been a time when only one man did, because he was the only guy who caught on, before he wrote his paper. But after people read the paper a lot of people understood the theory of relativity in some way or other, certainly more than twelve. On the other hand, I think I can safely say that nobody understands quantum mechanics. So do not take the lecture too seriously, feeling that you really have to understand in terms of some model what I am going to describe, but just relax and enjoy it. I am going to tell you what nature behaves like. If you will simply admit that maybe she does behave like this, you will find her a delightful, entrancing thing. Do not keep saying to yourself, if you can possibly avoid it, 'But how can it be like that?' because you will get 'down the drain', into a blind alley from which nobody has yet escaped. Nobody knows how it can be like that.

So then, let me describe to you the behaviour of electrons or of photons in their typical quantum mechanical way. I am going to do this by a mixture of analogy and contrast.

If I made it pure analogy we would fail; it must be by analogy and contrast with things which are familiar to you. So I make it by analogy and contrast, first to the behaviour of particles, for which I will use bullets, and second to the behaviour of waves, for which I will use water waves. What I am going to do is to invent a particular experiment and first tell you what the situation would be in that experiment using particles, then what you would expect to happen if waves were involved, and finally what happens when there are actually electrons or photons in the system. I will take just this one experiment, which has been designed to contain all of the mystery of quantum mechanics, to put you up against the paradoxes and mysteries and peculiarities of nature one hundred per cent. Any other situation in quantum mechanics, it turns out, can always be explained by saying, 'You remember the case of the experiment with the two holes? It's the same thing'. I am going to tell you about the experiment with the two holes. It does contain the general mystery; I am avoiding nothing; I am baring nature in her most elegant and difficult form.

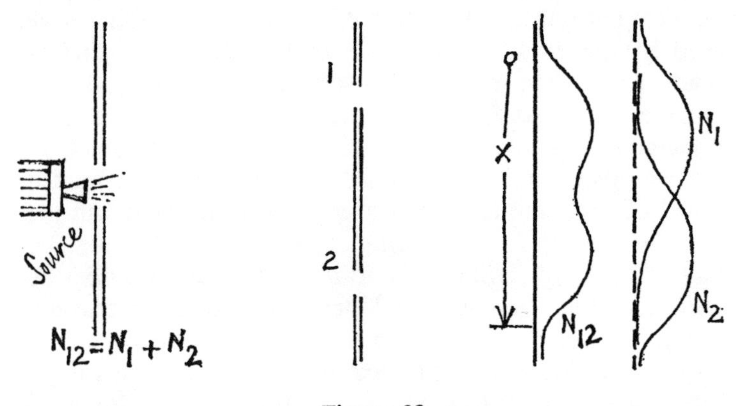

Figure 28

We start with bullets (fig. 28). Suppose that we have some source of bullets, a machine gun, and in front of it a plate with a hole for the bullets to come through, and this plate is armour plate. A long distance away we have a second plate which has two holes in it – that is the famous two-hole business. I am going to talk a lot about these holes, so I will call them hole No. 1 and hole No. 2. You can imagine round holes in three dimensions – the drawing is just a cross section. A long distance away again we have another screen which is just a backstop of some sort on which we can put in various places a detector, which in the case of bullets is a box of sand into which the bullets will be caught so that we can count them. I am going to do experiments in which I count how many bullets come into this detector or box of sand when the box is in different positions, and to describe that I will measure the distance of the box from somewhere, and call that distance 'x', and I will talk about what happens when you change 'x', which means only that you move the detector box up and down. First I would like to make a few modifications from real bullets, in three idealizations. The first is that the machine gun is very shaky and wobbly and the bullets go in various directions, not just exactly straight on; they can ricochet off the edges of the holes in the armour plate. Secondly, we should say, although this is not very important, that the bullets have all the same speed or energy. The most important idealization in which this situation differs from real bullets is that I want these bullets to be absolutely indestructible, so that what we find in the box is not pieces of lead, of some bullet that broke in half, but we get the whole bullet. Imagine indestructible bullets, or hard bullets and soft armour plate.

The first thing that we shall notice about bullets is that the things that arrive come in lumps. When the energy comes it is all in one bulletful, one bang. If you count the bullets, there are one, two, three, four bullets; the things come in lumps. They are of equal size, you suppose, in this case, and when a thing comes into the box it is either all in the box or it is not in the box. Moreover, if I put up two boxes I never get two bullets in the boxes at the same time, presuming that the gun is not going off too fast and I have enough time between them to see. Slow down the gun so it goes off very slowly, then look very quickly in the two boxes, and you will never get two bullets at the same time in the two boxes, because a bullet is a single identifiable lump.

Now what I am going to measure is how many bullets arrive on the average over a period of time. Say we wait an hour, and we count how many bullets are in the sand and average that. We take the number of bullets that arrive per hour, and we can call that the probability of arrival, because it just gives the chance that a bullet going through a slit arrives in the particular box. The number of bullets that arrive in the box will vary of course as I vary 'x'. On the diagram I have plotted horizontally the number of bullets that I get if I hold the box in each position for an hour. I shall get a curve that will look more or less like curve N_{12} because when the box is behind one of the holes it gets a lot of bullets, and if it is a little out of line it does not get as many, they have to bounce off the edges of the holes, and eventually the curve disappears. The curve looks like curve N_{12}, and the number that we get in an hour when both holes are open I will call N_{12}, which merely means the number which arrive through hole No. 1 and hole No. 2.

I must remind you that the number that I have plotted does not come in lumps. It can have any size it wants. It can be two and a half bullets in an hour, in spite of the fact that bullets come in lumps. All I mean by two and a half bullets per hour is that if you run for ten hours you will get twenty-five bullets, so on the average it is two and a half bullets. I am sure you are all familiar with the joke about the average family in the United States seeming to have two and a half children. It does not mean that there is a half child in any family – children come in lumps. Nevertheless, when you take the average number per family it can be any number whatsoever, and in the same way this number N_{12}, which is the number of bullets that arrive in the container per hour, on the average, need not be an integer. What we measure is the probability of arrival, which is a technical term for the average number that arrive in a given length of time.

Finally, if we analyse the curve N_{12} we can interpret it very nicely as the sum of two curves, one which will represent what I will call N_1, the number which will come if hole No. 2 is closed by another piece of armour plate in front, and N_2, the number which will come through hole No. 2 alone, if hole No. 1 is closed. We discover now a very important law, which is that the number that arrive with both holes open is the number that arrive by coming through hole No. 1, plus the number that come through hole No. 2. This proposition, the fact that all you have to do is to add these together, I call 'no interference'.

$$N_{12} = N_1 + N_2 \text{ (no interference).}$$

That is for bullets, and now we have done with bullets we begin again, this time with water waves (fig. 29). The

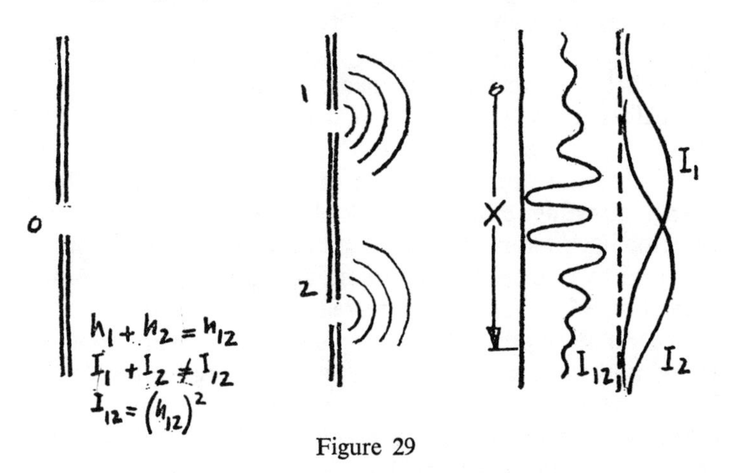

Figure 29

source is now a big mass of stuff which is being shaken up and down in the water. The armour plate becomes a long line of barges or jetties with a gap in the water in between. Perhaps it would be better to do it with ripples than with big ocean waves; it sounds more sensible. I wiggle my finger up and down to make waves, and I have a little piece of wood as a barrier with a hole for the ripples to come through. Then I have a second barrier with two holes, and finally a detector. What do I do with the detector? What the detector detects is how much the water is jiggling. For instance, I put a cork in the water and measure how it moves up and down, and what I am going to measure in fact is the energy of the agitation of the cork, which is exactly proportional to the energy carried by the waves. One other thing: the jiggling is made very regular and perfect so that the waves are all the same space from one another. One thing that is important for water waves is that the thing we are measuring can have any size at all. We are measuring the intensity of the waves, or the energy in the cork, and if the waves are very quiet, if my finger is only jiggling a little, then there will be very little motion of the cork. No matter how much it is, it is proportional. It can have any size; it does not come in lumps; it is not all there or nothing.

What we are going to measure is the intensity of the waves, or, to be precise, the energy generated by the waves at a point. What happens if we measure this intensity, which I will call 'I' to remind you that it is an intensity and not a number of particles of any kind? The curve I_{12}, that is when both holes are open, is shown in the diagram (fig. 29). It is an interesting, complicated looking curve. If we put the detector in different places we get an intensity which varies very rapidly in a peculiar manner. You are probably familiar with the reason for that. The reason is that the ripples as they come have crests and troughs spreading from hole No. 1, and they have crests and troughs spreading from hole No. 2. If we are at a place which is exactly in between the two holes, so that the two waves arrive at the same time, the crests will come on top of each other and there will be plenty of jiggling. We have a lot of jiggling right in dead centre. On the other hand if I move the detector to some point further from hole No. 2 than hole No. 1, it takes a little longer for the waves to come from 2 than from 1, and when a crest is arriving from 1 the crest has not quite reached there yet from hole 2, in fact it is a trough from 2, so that the water tries to move up and it tries to move down, from the influences of the waves coming from the two holes, and the net result is that it does not move at all, or practically not at all. So we have a low bump at that place. Then if it moves still further over we get enough delay so that crests come together from both holes, although one crest is in fact a whole wave behind, and so you get a big one again, then a small one, a big one, a small one . . . depending upon the way the crests and troughs 'interfere'. The word interference again is used in science in a funny way. We can have what we call constructive interference, as when both waves interfere to make the intensity stronger. The important thing is that I_{12} is not the same as I_1 plus I_2, and we say it shows constructive and destructive interference. We can find out what I_1 and I_2 look like by closing hole No. 2 to find I_1, and closing hole No. 1 to find I_2. The intensity that we get if one hole is closed is simply the waves from one hole, with no interference, and the curves are shown in fig. 2. You will notice that I_1 is the same as N_1, and I_2 the same as N_2 and yet I_{12} is quite different from N_{12}.

As a matter of fact, the mathematics of the curve I_{12} is rather interesting. What is true is that the height of the water, which we will call h, when both holes are open is equal to the height that you would get from No. 1 open, plus the height that you would get from No. 2 open. Thus, if it is a trough the height from No. 2 is negative and cancels out the height from No. 1. You can represent it by talking about the height of the water, but it turns out that the intensity in any case, for instance when both holes are open, is not the same as the height but is proportional to the square of the height. It is because of the fact that we are dealing with squares that we get these very interesting curves.

$$h_{12} = h_1 + h_2$$

but

$$I_{12} \neq I_1 + I_2 \text{ (Interference)}$$

$$I_{12} = (h_{12})^2,$$

$$I_1 = (h_1)^2$$

$$I_2 = (h_2)^2$$

That was water. Now we start again, this time with electrons (fig. 30).

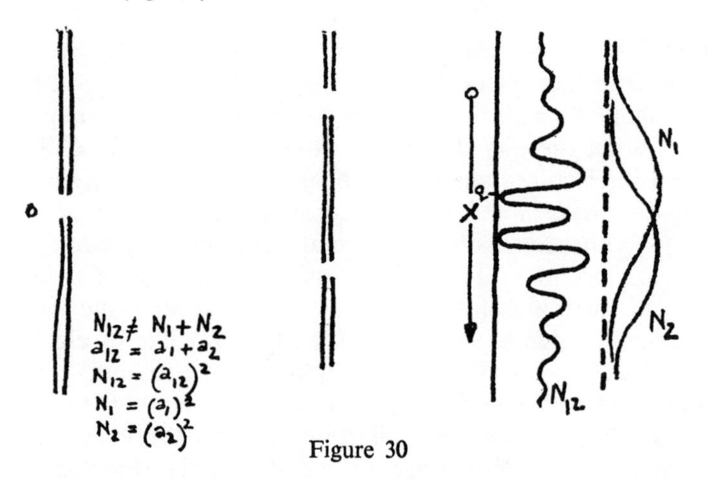

Figure 30

The source is a filament, the barriers tungsten plates, these are holes in the tungsten plate, and for a detector we have any electrical system which is sufficiently sensitive to pick up the charge of an electron arriving with whatever energy the source has. If you would prefer it, we could use photons with black paper instead of the tungsten plate – in fact black paper is not very good because the fibres do not make sharp holes, so we would have to have something better – and for a detector a photo-multiplier capable of detecting the individual photons arriving. What happens with either case? I will discuss only the electron case, since the case with photons is exactly the same.

First, what we receive in the electrical detector, with a sufficiently powerful amplifier behind it, are clicks, lumps, absolute lumps. When the click comes it is a certain size, and the size is always the same. If you turn the source weaker the clicks come further apart, but it is the same sized click. If you turn it up they come so fast that they jam the amplifier. You have to turn it down enough so that there are not too many clicks for the machinery that you are using for the detector. Next, if you put another detector in a different place and listen to both of them you will never get two clicks at the same time, at least if the source is weak enough and the precision with which you measure the time is good enough. If you cut down the intensity of the source so that the electrons come few and far between, they never give a click in both detectors at once. That means that the thing which is coming comes in lumps – it has a definite size, and it only comes to one place at a time. Right, so electrons, or photons, come in lumps. Therefore what we can do is the same thing as we did for bullets: we can measure the probability of arrival. What we do is hold the detector in various places – actually if we wanted to although it is expensive, we could put detectors all over at the same time and make the whole curve simultaneously – but we hold the detector in each place, say for an hour, and we measure at the end of the hour how many electrons came, and we average it. What do we get for the number of electrons that arrive? The same kind of N_{12} as with bullets? Figure 30 shows what we get for N_{12}, that is what we get with both holes open. That is the phenomenon of nature, that she produces the curve which is the same as you would get for the interference of waves. She produces this curve for what? Not for the energy in a wave but for the probability of arrival of one of these lumps.

The mathematics is simple. You change I to N, so you have to change h to something else, which is new – it is not the height of anything – so we invent an 'a', which we call a probability amplitude, because we do not know what it means. In this case a_1 is the probability amplitude to arrive from hole No. 1, and a_2 the probability amplitude to arrive from hole No. 2. To get the total probability amplitude to arrive you add the two together and square it. This is a direct imitation of what happens with the waves, because we have to get the same curve out so we use the same mathematics.

I should check on one point though, about the interference. I did not say what happens if we close one of the holes. Let us try to analyse this interesting curve by presuming that the electrons came through one hole or through the other. We close one hole, and measure how many come through hole No. 1, and we get the simple curve N_1. Or we can close the other hole and measure how many come through hole No. 2, and we get the N_2 curve. But these two added together do not give the same as N_1+N_2; it does show interference. In fact the mathematics is given by this funny formula that the probability of arrival is the square of an amplitude which itself is the sum of two pieces, $N_{12} = (a_1+a_2)^2$. The question is how it can come about that when the electrons go through hole No. 1 they will be distributed one way, when they go through hole No. 2 they will be distributed another way, and yet when both holes are open you do not get the sum of the two. For instance, if I hold the detector at the point q with both holes open I get practically nothing, yet if I close one of the holes I get plenty, and if I close the other hole I get something. I leave both holes open and I get nothing; I let them come through both holes and they do not come any more. Or take the point at the centre; you can show that that is higher than the sum of the two single hole curves. You might think that if you were clever enough you could argue that they have some way of going around through the holes back and forth, or they do something complicated, or one splits in half and goes through the two holes, or something similar, in order to explain this phenomenon. Nobody, however, has succeeded in producing an explanation that is satisfactory, because the mathematics in the end are so very simple, the curve is so very simple (fig. 30).

I will summarize, then, by saying that electrons arrive in lumps, like particles, but the probability of arrival of these lumps is determined as the intensity of waves would be. It is in this sense that the electron behaves sometimes like a particle and sometimes like a wave. It behaves in two different ways at the same time (fig. 31).

That is all there is to say. I could give a mathematical description to figure out the probability of arrival of elec-

TABLE

BULLETS	WATER WAVES	ELECTRONS (PHOTONS)
COME IN LUMPS	CAN HAVE ANY SIZE	COME IN LUMPS
MEASURE PROBABILITY OF ARRIVAL	MEASURE INTENSITY OF WAVES	MEASURE PROBABILITY OF ARRIVAL
$N_{12} = N_1 + N_2$	$I_{12} \neq I_1 + I_2$	$N_{12} \neq N_1 + N_2$
NO INTERFERENCE	SHOWS INTERFERENCE	SHOWS INTERFERENCE

Figure 31

trons under any circumstances, and that would in principle be the end of the lecture – except that there are a number of subtleties involved in the fact that nature works this way. There are a number of peculiar things, and I would like to discuss those peculiarities because they may not be self-evident at this point.

To discuss the subtleties, we begin by discussing a proposition which we would have thought reasonable, since these things are lumps. Since what comes is always one complete lump, in this case an electron, it is obviously reasonable to assume that either an electron goes through hole No. 1 or it goes through hole No. 2. It seems very obvious that it cannot do anything else if it is a lump. I am going to discuss this proposition, so I have to give it a name; I will call it 'proposition A'.

Proposition A.

Either an electron goes through hole N°1 or it goes through hole N°2.

Now we have already discussed a little what happens with proposition A. If it were true that an electron either goes through hole No. 1 or through hole No. 2, then the total number to arrive would have to be analysable as the sum of two contributions. The total number which arrive will be the number that come via hole 1, plus the number that come via hole 2. Since the resulting curve cannot be easily analysed as the sum of two pieces in such a nice manner, and since the experiments which determine how many would arrive if only one hole or the other were open do not give the result that the total is the sum of the two parts, it is obvious that we should conclude that this proposition is false. If it is not true that the electron either comes through hole No. 1 or hole No. 2, maybe it divides itself in half temporarily or something. So proposition A is false. That is logic. Unfortunately, or otherwise, we can test logic by experiment. We have to find out whether it is true or not that the electrons come through either hole 1 or hole 2, or maybe they go round through both holes and get temporarily split up, or something.

All we have to do is watch them. And to watch them we need light. So we put behind the holes a source of very intense light. Light is scattered by electrons, bounced off them, so if the light is strong enough you can see electrons as they go by. We stand back, then, and we look to see whether when an electron is counted we see, or have seen the moment before the electron is counted, a flash behind hole 1 or a flash behind hole 2, or maybe a sort of half flash in each place at the same time. We are going to find out now how it goes, by looking. We turn on the light and look, and lo, we discover that every time there is a count at the detector we see either a flash behind No. 1 or a flash behind No. 2. What we find is that the electron comes one hundred per cent, complete, through hole 1 or through hole 2 – when we look. A paradox!

Let us squeeze nature into some kind of a difficulty here. I will show you what we are going to do. We are going to keep the light on and we are going to watch and count how many electrons come through. We will make two columns, one for hole No. 1 and one for hole No. 2, and as each electron arrives at the detector we will note in the appropriate column which hole it came through. What does the column for hole No. 1 look like when we add it all together for different positions of the detector? If I watch behind hole No. 1 what do I see? I see the curve N_1 (fig. 30). That column is distributed just as we thought when we closed hole 2, much the same way whether we are looking or not. If we close hole 2 we get the same distribution in those that arrive as if we were watching hole No. 1; likewise the number that have arrived via hole No. 2 is also a simple curve N_2. Now look, the total number which arrive *has to be* the total number. It has to be the sum of the number N_1 plus the number N_2; because each one that comes through has been checked off in either column 1 or column 2. The total number which arrive *absolutely has to be* the sum of these two. It has to be distributed as N_1+N_2. But I said it was distributed as the curve N_{12}. No, it is distributed as N_1+N_2. It really is, of course; it has to be and it is. If we mark with a prime the results when a light is lit, then we find that N_1', is practically the same as N_1, without the light, and N_2' is almost the same as N_2. But the number N_{12}', that we see when the light is on and both holes are open *is* equal to the number that we see through hole 1 plus the number that we see through hole 2. This is the result that we get when the light is on. We get a different answer whether I turn on the light or not. If I have the light turned on, the distribution is the curve N_1+N_2. If I turn off the light, the distribution is N_{12}. Turn on the light and it is N_1+N_2 again. So you see, nature has squeezed out! We could say, then, that the light affects the result. If the light is on you get a different answer from that when the light is off. You can say too that light affects the behaviour of electrons. If you talk about the motion of the electrons through the experiment, which is a little inaccurate, you can say that the light affects the motion, so that those which might have arrived at the maximum have somehow been deviated or kicked by the light and arrive at the minimum instead, thus smoothing the curve to produce the simple N_1+N_2 curve.

Electrons are very delicate. When you are looking at a baseball and you shine a light on it, it does not make any difference, the baseball still goes the same way. But when you shine a light on an electron it knocks him about a bit, and instead of doing one thing he does another, because you have turned the light on and it is so strong. Suppose we try turning it weaker and weaker, until it is very dim, then use very careful detectors that can see very dim lights, and look with a dim light. As the light gets dimmer and dimmer you cannot expect the very very weak light to affect the electron so completely as to change the pattern a hundred per cent from N_{12} to N_1+N_2. As the light gets weaker and weaker, somehow it should get more and more like no light at all. How then does one curve turn into another? But of course light is not like a wave of water. Light also comes in particle-like character, called photons, and as you turn down the intensity of the light you are not turning down the effect, you are turning down the number of photons that are coming out of the source. As I turn down the light I am getting fewer and fewer photons. The least I can scatter from an electron is one photon, and if I have too few photons sometimes the electron will get through when there is no photon coming by, in which case I will not see it. A very weak light, therefore, does not mean a small disturbance, it just means a few photons. The result is that with a very weak light I have to invent a third column under the title 'didn't see'. When the light is very strong there are few in there, and when the light is very weak most of them end in there. So there are three columns, hole 1, hole 2, and didn't see. You can guess what happens. The ones I do see are distributed according to the curve N_1+N_2. The ones I do not see are distributed as the curve N_{12}. As I turn the light weaker and weaker I see less and less and a greater and greater fraction are not seen. The actual curve in any case is a mixture of the two curves, so as the light gets weaker it gets more and more like N_{12} in a continuous fashion.

I am not able here to discuss a large number of different ways which you might suggest to find out which hole the electron went through. It always turns out, however, that it is impossible to arrange the light in any way so that you can tell through which hole the thing is going without disturbing the pattern of arrival of the electrons, without destroying the interference. Not only light, but anything else – whatever you use, in principle it is impossible to do it. You can, if you want, invent many ways to tell which hole the electron is going through, and then it turns out that it is going through one or the other. But if you try to make that instrument so that at the same time it does not disturb the motion of the electron, then what happens is that you can no longer tell which hole it goes through and you get the complicated result again.

Heisenberg noticed, when he discovered the laws of quantum mechanics, that the new laws of nature that he had discovered could only be consistent if there were some basic limitation to our experimental abilities that had not been previously recognized. In other words, you cannot experimentally be as delicate as you wish. Heisenberg proposed his uncertainty principle which, stated in terms of our own experiment, is the following. (He stated it in another way, but they are exactly equivalent, and you can get from one to the other.) 'It is impossible to design any apparatus whatsoever to determine through which hole the electron passes that will not at the same time disturb the electron enough to destroy the interference pattern'. No one has found a way around this. I am sure you are itching with inventions of methods of detecting which hole the electron went through; but if each one of them is analysed carefully you will find out that there is something the matter with it. You may think you could do it without disturbing the electron, but it turns out there is always something the matter, and you can always account for the difference in the patterns by the disturbance of the instruments used to determine through which hole the electron comes.

This is a basic characteristic of nature, and tells us

something about everything. If a new particle is found tomorrow, the kaon – actually the kaon has already been found, but to give it a name let us call it that – and I use kaons to interact with electrons to determine which hole the electron is going through, I already know, ahead of time – I hope – enough about the behaviour of a new particle to say that it cannot be of such a type that I could tell through which hole the electron would go without at the same time producing a disturbance on the electron and changing the pattern from interference to no interference. The uncertainty principle can therefore be used as a general principle to guess ahead at many of the characteristics of unknown objects. They are limited in their likely character.

Let us return to our proposition A – 'Electrons must go either through one hole or another'. Is it true or not? Physicists have a way of avoiding the pitfalls which exist. They make their rules of thinking as follows. If you have an apparatus which is capable of telling which hole the electron goes through (and you *can* have such an apparatus), then you can say that it either goes through one hole or the other. It does; it always is going through one hole or the other – when you look. But when you have no apparatus to determine through which hole the thing goes, then you cannot say that it either goes through one hole or the other. (You can always *say* it – provided you stop thinking immediately and make no deductions from it. Physicists prefer not to say it, rather than to stop thinking at the moment.) To conclude that it goes either through one hole or the other when you are not looking is to produce an error in prediction. That is the logical tight-rope on which we have to walk if we wish to interpret nature.

This proposition that I am talking about is general. It is not just for two holes, but is a general proposition which can be stated this way. The probability of any event in an ideal experiment – that is just an experiment in which everything is specified as well as it can be – is the square of something, which in this case I have called 'a', the probability amplitude. When an event can occur in several alternative ways, the probability amplitude, this 'a' number, is the sum of the 'a's for each of the various alternatives. If an experiment is performed which is capable of determining which alternative is taken, the probability of the event is changed; it is then the sum of the probabilities for each alternative. That is, you lose the interference.

The question now is, how does it really work? What machinery is actually producing this thing? Nobody knows any machinery. Nobody can give you a deeper explanation of this phenomenon than I have given; that is, a description of it. They can give you a wider explanation, in the sense that they can do more examples to show how it is impossible to tell which hole the electron goes through and not at the same time destroy the interference pattern. They can give a wider class of experiments than just the two slit interference experiment. But that is just repeating the same thing to drive it in. It is not any deeper; it is only wider. The mathematics can be made more precise; you can mention that they are complex numbers instead of real numbers, and a couple of other minor points which have nothing to do with the main idea. But the deep mystery is what I have described, and no one can go any deeper today.

What we have calculated so far is the probability of arrival of an electron. The question is whether there is any way to determine where an individual electron really arrives? Of course we are not averse to using the theory of probability, that is calculating odds, when a situation is very complicated. We throw up a dice into the air, and with the various resistances, and atoms, and all the complicated business, we are perfectly willing to allow that we do not know enough details to make a definite prediction; so we calculate the odds that the thing will come this way or that way. But here what we are proposing, is it not, is that there is probability all the way back: that in the fundamental laws of physics there are odds.

Suppose that I have an experiment so set up that with the light out I get the interference situation. Then I say that even with the light on I cannot predict through which hole an electron will go. I only know that each time I look it will be one hole or the other; there is no way to predict ahead of time which hole it will be. The future, in other words, is unpredictable. It is impossible to predict in any way, from any information ahead of time, through which hole the thing will go, or which hole it will be seen behind. That means that physics has, in a way, given up, if the original purpose was – and everybody thought it was – to know enough so that given the circumstances we can predict what will happen next. Here are the circumstances: electron source, strong light source, tungsten plate with two holes: tell me, behind which hole shall I see the electron? One theory is that the reason you cannot tell through which hole you are going to see the electron is that it is determined by some very complicated things back at the source: it has internal wheels, internal gears, and so forth, to determine which hole it goes through; it is fifty-fifty probability, because, like a die, it is set at random; physics is incomplete, and if we get a complete enough physics then we shall be able to predict through which hole it goes. That is called the hidden variable theory. That theory cannot be true; it is not due to lack of detailed knowledge that we cannot make a prediction.

I said that if I did not turn on the light I should get the interference pattern. If I have a circumstance in which I get that interference pattern, then it is impossible to analyse it in terms of saying it goes through hole 1 or hole 2, because that interference curve is so simple, mathematically a completely different thing from the contribution of the two other curves as probabilities. If it had been possible for us to determine through which hole the electron was going to go if we had the light on, then whether we have the light on or off is nothing to do with it. Whatever gears there are at the source, which we observed, and which permitted us to tell whether the thing was going to go through 1 or 2, we could have observed with the light off, and therefore we could have told with the light off through which hole each electron was going to go. But if we could do this, the resulting curve would have to be represented as the sum of those that go through hole 1 and those that go through hole 2, and it is not. It must then be impossible to have any information ahead of time about which hole the electron is going to go through, whether the light is on or off, in any circumstance when the experiment is set up so that it can produce the interference with the light off. It is not our ignorance of the internal gears, of the internal complications, that makes nature appear to have probability in it. It seems to be somehow intrinsic. Someone has said it this way – 'Nature herself does not even know which way the electron is going to go'.

A philosopher once said 'It is necessary for the very existence of science that the same conditions always produce the same results'. Well, they do not. You set up the circumstances, with the same conditions every time, and you cannot predict behind which hole you will see the electron. Yet science goes on in spite of it – although the same conditions do not always produce the same results. That makes us unhappy, that we cannot predict exactly what will happen. Incidentally, you could think up a circumstance in which it is very dangerous and serious, and man *must* know, and still you cannot predict. For instance we could cook up – we'd better not, but we could – a scheme by which we set up a photo cell, and one electron to go through, and if we see it behind hole No. 1 we set off the atomic bomb and start World War III, whereas if we see it behind hole No. 2 we make peace feelers and delay the war a little longer. Then the future of man would be dependent on something which no amount of science can predict. The future is unpredictable.

What is necessary 'for the very existence of science', and what the characteristics of nature are, are not to be determined by pompous preconditions, they are determined always by the material with which we work, by nature herself. We look, and we see what we find, and we cannot say ahead of time successfully what it is going to look like. The most reasonable possibilities often turn out not to be the situation. If science is to progress, what we need is the ability to experiment, honesty in reporting results – the results must be reported without somebody saying what they would like the results to have been – and finally – an important thing – the intelligence to interpret the results. An important point about this intelligence is that it should not be sure ahead of time what must be. It can be prejudiced, and say 'That is very unlikely; I don't like that'. Prejudice is different from absolute certainty. I do not mean absolute prejudice – just bias. As long as you are only biased it does not make any difference, because if your bias is wrong a perpetual accumulation of experiments will perpetually annoy you until they cannot be disregarded any longer. They can only be disregarded if you are absolutely sure ahead of time of some precondition that science has to have. In fact it is necessary for the very existence of science that minds exist which do not allow that nature must satisfy some preconceived conditions, like those of our philosopher.

7

Seeking New Laws

What I want to talk about in this lecture is not, strictly speaking, the character of physical law. One might imagine at least that one is talking about nature when one is talking about the character of physical law; but I do not want to talk about nature, but rather about how we stand relative to nature now. I want to tell you what we think we know, what there is to guess, and how one goes about guessing. Someone suggested that it would be ideal if, as I went along, I would slowly explain how to guess a law, and then end by creating a new law for you. I do not know whether I shall be able to do that.

First I want to tell you what the present situation is, what it is that we know about physics. You may think that I have told you everything already, because in the lectures I have told you all the great principles that are known. But the principles must be principles about *something*; the principle of the conservation of energy relates to the energy of *something*, and the quantum mechanical laws are quantum mechanical laws about *something* – and all these principles added together still do not tell us what the content is of the nature that we are talking about. I will tell you a little, then, about the stuff on which all of these principles are supposed to have been working.

First of all there is matter – and, remarkably enough, all matter is the same. The matter of which the stars are made is known to be the same as the matter on the earth. The character of the light that is emitted by those stars gives a kind of fingerprint by which we can tell that there are the same kinds of atoms there as on the earth. The same kinds of atoms appear to be in living creatures as in non-living creatures; frogs are made of the same 'goup' as rocks, only in different arrangements. So that makes our problem simpler; we have nothing but atoms, all the same, everywhere.

The atoms all seem to be made from the same general constitution. They have a nucleus, and around the nucleus there are electrons. We can make a list of the parts of the world that we think we know about (fig. 32).

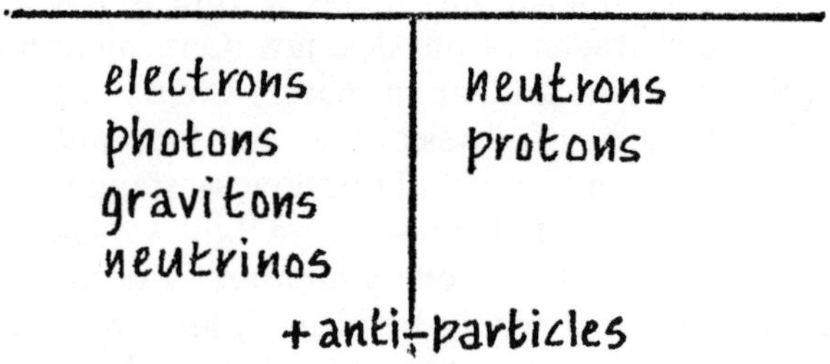

electrons	neutrons
photons	protons
gravitons	
neutrinos	

+ anti-particles

Figure 32

First there are the electrons, which are the particles on the outside of the atom. Then there are the nuclei; but those are understood today as being themselves made up of two other things which are called neutrons and protons – two particles. We have to see the stars, and see the atoms, and they emit light, and the light itself is described by particles which are called photons. In the beginning we spoke about gravitation; and if the quantum theory is right, then the gravitation should have some kind of waves which behave like particles too, and these are called gravitons. If you do not believe in that, just call it gravity. Finally, I did mention what is called beta-decay, in which a neutron can disintegrate into a proton, an electron and a neutrino – or really an anti-neutrino; there is another particle, a neutrino. In addition to all the particles I have listed there are of course all the anti-particles; that is just a quick statement that takes care of doubling the number of particles, but there is no complication.

With these particles that I have listed, all of the low energy phenomena, in fact all ordinary phenomena that happen everywhere in the Universe, so far as we know, can be explained. There are exceptions, when here and there some very high energy particle does something, and in the laboratory we have been able to do some peculiar things. But if we leave out these special cases, all ordinary phenomena can be explained by the actions and the motions of particles. For example, life itself is supposedly understandable in principle from the movements of atoms, and those atoms are made out of neutrons, protons and electrons. I must immediately say that when we state that we understand it in principle, we only mean that we think that, if we could figure everything out, we would find that there is nothing new in physics which needs to be discovered in order to understand the phenomena of life. Another instance, the fact that the stars emit energy, solar energy or stellar energy, is presumably also understood in terms of nuclear reactions among these particles. All kinds of details of the way atoms behave are accurately described with this kind of model, at least as far as we know at present. In fact, I can say that in the range of phenomena today, so far as I know there are no phenomena that we are sure cannot be explained this way, or even that there is deep mystery about.

This was not always possible. There is, for instance, a phenomenon called super-conductivity, which means that metals conduct electricity without resistance at low temperatures. It was not at first obvious that this was a consequence of the known laws. Now that it has been thought through carefully enough, it is seen in fact to be fully explainable in terms of our present knowledge. There are other phenomena, such as extra-sensory perception, which cannot be explained by our knowledge of physics. However, that phenomenon has not been well established, and we cannot guarantee that it is there. If it could be demonstrated, of course, that would prove that physics is incomplete, and it is therefore extremely interesting to physicists whether it is right or wrong. Many experiments exist which show that it does not work. The same goes for astrological influences. If it were true that the stars could affect the day that it was good to go to the dentist – in America we have that kind of astrology – then physics theory would be proved wrong, because there is no mechanism understandable in principle from the behaviour of particles which would make this work. That is the reason that there is some scepticism among scientists with regard to those ideas.

On the other hand, in the case of hypnotism, at first it looked as though that also would be impossible, when it was described incompletely. Now that it is known better it is realized that it is not absolutely impossible that hypnosis could occur through normal physiological, though as yet unknown, processes; it does not obviously require some special new kind of force.

Today, although our theory of what goes on outside the nucleus of the atom seems precise and complete enough, in the sense that given enough time we can calculate anything as accurately as it can be measured, it turns out that the forces between neutrons and protons, which constitute the nucleus, are not so completely known, and are not understood at all well. What I mean is that we do not today understand the forces between neutrons and protons to the extent that if you wanted me to, and gave me enough time and computers, I could calculate exactly the energy levels of carbons, or something like that. We do not know enough. Although we can do the corresponding thing for the energy levels of the outside electrons of the atom, we cannot for the nucleus, since the nuclear forces are still not understood very well.

In order to find out more about this, experimenters have gone on to study phenomena at very high energy. They hit neutrons and protons together at very high energy to produce peculiar things, and by studying these peculiar things we hope to understand better the forces between neutrons and protons. Pandora's box has been opened by these experiments! Although all we really wanted was to get a better idea of the forces between neutrons and protons, when we hit these things together hard we discovered that there are more particles in the world. In fact more than four dozen other particles have been dredged up in an attempt to understand these forces; we will put these four dozen others into the neutron/proton column (fig. 33), because they inter-

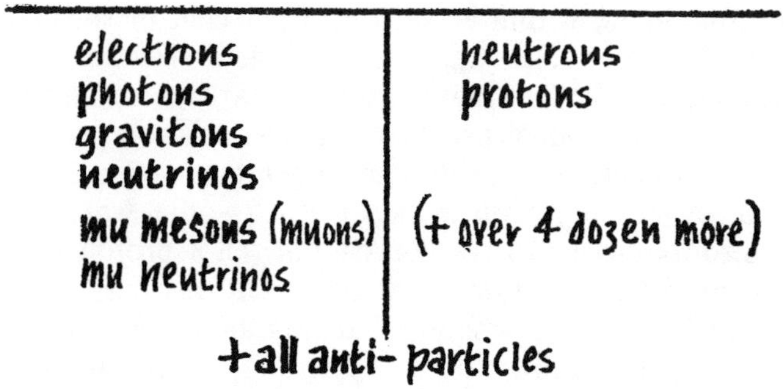

electrons	neutrons
photons	protons
gravitons	
neutrinos	
mu mesons (muons)	(+ over 4 dozen more)
mu neutrinos	

+ all anti-particles

Figure 33

act with neutrons and protons, and have something to do with the forces between them. In addition to that, while the dredge was digging up all this mud it picked up a couple of pieces that are irrelevant to the problem of nuclear forces. One of them is called a mu meson, or muon, and the other is a neutrino which goes with it. There are two kinds of neutrino, one which goes with the electron and one which goes with the mu meson. Incidentally, most amazingly, all the laws of the muon and its neutrino are now known, as far as we can tell experimentally, and the law is that they behave in precisely the same way as the electron and its neutrino, except that the mass of the mu meson is 207 times heavier than the electron; but that is the only difference known between those objects, which is rather curious. Four dozen other particles is a frightening array – plus the anti-particles. They have various names, mesons, pions, kaons, lambda, sigma . . . it does not make any difference . . . with four dozen particles there are going to be a lot of names!

But it turns out that these particles come in families, which helps us a little. Actually some of these so-called particles last such a short time that there are debates about whether it is in fact possible to define their very existence, but I will not enter into that debate.

In order to illustrate the family idea, I will take the cases of a neutron and a proton. The neutron and the proton have the same mass, within a tenth of a per cent or so. One is 1,836, the other 1,839 times as heavy as an electron. More remarkable is the fact that for the nuclear forces, the strong forces inside the nucleus, the force between two protons is the same as between a proton and a neutron, and is the same again between a neutron and a neutron. In other words, from the strong nuclear forces you cannot tell a proton from a neutron. So it is a symmetry law; neutrons may be substituted for protons without changing anything – provided you are only talking about the strong forces. But if you change a neutron for a proton you have a terrific difference, because the proton carries an electrical charge and the neutron does not. By electrical measurement you can immediately see the difference between a proton and a neutron, so this symmetry, that you can replace one by the other, is what we call an approximate symmetry. It is right for the strong interactions of nuclear forces, but it is not right in any deep sense of nature, because it does not work for electricity. This is called a partial symmetry, and we have to struggle with these partial symmetries.

Now that the families have been extended, it turns out that substitutions of the type of neutron for proton can be extended over a wider range of particles. But the accuracy is still lower. The statement that neutrons can always be substituted for protons is only approximate – it is not true for electricity – but the wider substitutions which have been found possible give a still poorer symmetry. However, these partial symmetries have helped to gather the particles into families and thus to locate places where particles are missing and to help to discover new ones.

This kind of game, of roughly guessing at family relationships and so on, is illustrative of the kind of preliminary sparring which one does with nature before really discovering some deep and fundamental law. Examples are very important in the previous history of science. For instance, Mendeleev's* discovery of the periodic table of the elements is analogous to this game. It is the first step; but the complete description of the reason for the atomic table came much later, with atomic theory. In the same way, organization of the knowledge of nuclear levels was made by Maria Mayer and Jensen† in what they called the shell model of nuclei some years ago. Physics is in an analogous game, in which a reduction of the complexity is made by some approximate guesses.

In addition to these particles we have all the principles that we were talking about before, the principles of symmetry, of relativity, and that things must behave quantum mechanically; and, combining that with relativity, that all conservation laws must be local.

If we put all these principles together, we discover that there are too many. They are inconsistent with each other. It seems that if we take quantum mechanics, plus relativity, plus the proposition that everything has to be local, plus a number of tacit assumptions, we get inconsistency, because we get infinity for various things when we calculate them, and if we get infinity how can we ever say that this agrees with nature? An example of these tacit assumptions which I mentioned, about which we are too prejudiced to understand the real significance, is such a proposition as the following. If you calculate the chance for every possibility – say it is 50% probability this will happen, 25% that will happen, etc., it should add up to 1. We think that if you

*Dimitri Ivanovitch Mendeleev, 1834–1907, Russian chemist.

†Maria Mayer, American physicist, Nobel Prize 1963, Professor of Physics at University of California since 1960. Hans Daniel Jensen, German physicist, Nobel Prize, 1963. Director of Institute for Theoretical Physics at Heidelberg since 1949.

add all the alternatives you should get 100% probability. That seems reasonable, but reasonable things are where the trouble always is. Another such proposition is that the energy of something must always be positive – it cannot be negative. Another proposition which is probably added in before we get inconsistency is what is called causality, which is something like the idea that effects cannot precede their causes. Actually no one has made a model in which you disregard the proposition about the probability, or you disregard the causality, which is also consistent with quantum mechanics, relativity, locality and so on. So we really do not know exactly what it is that we are assuming that gives us the difficulty producing infinities. A nice problem! However, it turns out that it is possible to sweep the infinities under the rug, by a certain crude skill, and temporarily we are able to keep on calculating.

O.K., that is the present situation. Now I am going to discuss how we would look for a new law.

In general we look for a new law by the following process. First we guess it. Then we compute the consequences of the guess to see what would be implied if this law that we guessed is right. Then we compare the result of the computation to nature, with experiment or experience, compare it directly with observation, to see if it works. If it disagrees with experiment it is wrong. In that simple statement is the key to science. It does not make any difference how beautiful your guess is. It does not make any difference how smart you are, who made the guess, or what his name is – if it disagrees with experiment it is wrong. That is all there is to it. It is true that one has to check a little to make sure that it is wrong, because whoever did the experiment may have reported incorrectly, or there may have been some feature in the experiment that was not noticed, some dirt or something; or the man who computed the consequences, even though it may have been the one who made the guesses, could have made some mistake in the analysis. These are obvious remarks, so when I say if it disagrees with experiment it is wrong, I mean after the experiment has been checked, the calculations have been checked, and the thing has been rubbed back and forth a few times to make sure that the consequences are logical consequences from the guess, and that in fact it disagrees with a very carefully checked experiment.

This will give you a somewhat wrong impression of science. It suggests that we keep on guessing possibilities and comparing them with experiment, and this is to put experiment into a rather weak position. In fact experimenters have a certain individual character. They like to do experiments even if nobody has guessed yet, and they very often do their experiments in a region in which people know the theorist has not made any guesses. For instance, we may know a great many laws, but do not know whether they really work at high energy, because it is just a good guess that they work at high energy. Experimenters have tried experiments at higher energy, and in fact every once in a while experiment produces trouble; that is, it produces a discovery that one of the things we thought right is wrong. In this way experiment can produce unexpected results, and that starts us guessing again. One instance of an unexpected result is the mu meson and its neutrino, which was not guessed by anybody at all before it was discovered, and even today nobody yet has any method of guessing by which this would be a natural result.

You can see, of course, that with this method we can attempt to disprove any definite theory. If we have a definite theory, a real guess, from which we can conveniently compute consequences which can be compared with experiment, then in principle we can get rid of any theory. There is always the possibility of proving any definite theory wrong; but notice that we can never prove it right. Suppose that you invent a good guess, calculate the consequences, and discover every time that the consequences you have calculated agree with experiment. The theory is then right? No, it is simply not proved wrong. In the future you could compute a wider range of consequences, there could be a wider range of experiments, and you might then discover that the thing is wrong. That is why laws like Newton's laws for the motion of planets last such a long time. He guessed the law of gravitation, calculated all kinds of consequences for the system and so on, compared them with experiment – and it took several hundred years before the slight error of the motion of Mercury was observed. During all that time the theory had not been proved wrong, and could be taken temporarily to be right. But it could never be proved right, because tomorrow's experiment might succeed in proving wrong what you thought was right. We never are definitely right, we can only be sure we are wrong. However, it is rather remarkable how we can have some ideas which will last so long.

One of the ways of stopping science would be only to do experiments in the region where you know the law. But experimenters search most diligently, and with the greatest effort, in exactly those places where it seems most likely that we can prove our theories wrong. In other words we are trying to prove ourselves wrong as quickly as possible, because only in that way can we find progress. For example, today among ordinary low energy phenomena we do not know where to look for trouble, we think everything is all right, and so there is no particular big programme looking for trouble in nuclear reactions, or in super-conductivity. In these lectures I am concentrating on discovering fundamental laws. The whole range of physics, which is interesting, includes also an understanding at another level of these phenomena like super-conductivity and nuclear reactions, in terms of the fundamental laws. But I am talking now about discovering trouble, something wrong with the fundamental laws, and since among low energy phenomena nobody knows where to look, all the experiments today in this field of finding out a new law, are of high energy.

Another thing I must point out is that you cannot prove a vague theory wrong. If the guess that you make is poorly expressed and rather vague, and the method that you use for figuring out the consequences is a little vague – you are not sure, and you say, 'I think everything's right because it's all due to so and so, and such and such do this and that more or less, and I can sort of explain how this works . . .', then you see that this theory is good, because it cannot be proved wrong! Also if the process of computing the consequences is indefinite, then with a little skill any experimental results can be made to look like the expected consequences. You are probably familiar with that in other fields. 'A' hates his mother. The reason is, of course, because she did not caress him or love him enough when he was a child. But if you investigate you find out that as a matter of fact she did love him very much, and everything was all right. Well then, it was because she was over-indulgent when he was a child! By having a vague theory it is possible to get either result. The cure for this one is the following. If it were possible to state exactly, ahead of time, how much love is not enough, and how much love is over-indulgent, then there would be a perfectly legitimate theory against which you could make tests. It is usually said when this is pointed out, 'When you are dealing with psychological matters things can't be defined so precisely'. Yes, but then you cannot claim to know anything about it.

You will be horrified to hear that we have examples in physics of exactly the same kind. We have these approximate symmetries, which work something like this. You have an approximate symmetry, so you calculate a set of consequences supposing it to be perfect. When compared with experiment, it does not agree. Of course – the symmetry you are supposed to expect is approximate, so if the agreement is pretty good you say, 'Nice!', while if the agreement is very poor you say, 'Well, this particular thing must be especially sensitive to the failure of the symmetry'. Now you may laugh, but we have to make progress in that way. When a subject is first new, and these particles are new to us, this jockeying around, this 'feeling' way of guessing at the results, is the beginning of any science. The same thing is true of the symmetry proposition in physics as is true of psychology, so do not laugh too hard. It is necessary in the beginning to be very careful. It is easy to fall into the deep end by this kind of vague theory. It is hard to prove it wrong, and it takes a certain skill and experience not to walk off the plank in the game.

In this process of guessing, computing consequences, and comparing with experiment, we can get stuck at various stages. We may get stuck in the guessing stage, when we have no ideas. Or we may get stuck in the computing stage. For example, Yukawa* guessed an idea for the nuclear forces in 1934, but nobody could compute the consequences because the mathematics was too difficult, and so they could not compare his idea with experiment. The theories remained for a long time, until we discovered all these extra particles which were not contemplated by Yukawa, and therefore it is undoubtedly not as simple as the way Yukawa did it. Another place where you can get stuck is at the experimental end. For example, the quantum theory of gravitation is going very slowly, if at all, because all the experiments that you can do never involve quantum mechanics and gravitation at the same time. The gravity force is too weak compared with the electrical force.

Because I am a theoretical physicist, and more delighted with this end of the problem, I want now to concentrate on how you make the guesses.

As I said before, it is not of any importance where the guess comes from; it is only important that it should agree with experiment, and that it should be as definite as possible. 'Then', you say, 'that is very simple. You set up a machine, a great computing machine, which has a random wheel in it that makes a succession of guesses, and each time it guesses a hypothesis about how nature should work it computes immediately the consequences, and makes a comparison with a list of experimental results it has at the other end'. In other words, guessing is a dumb man's job. Actually it is quite the opposite, and I will try to explain why.

The first problem is how to start. You say, 'Well I'd start off with all the known principles'. But all the principles

*Hideki Yukawa, Japanese physicist. Director of Research Institute for Fundamental Physics at Kyoto. Nobel Prize 1949.

that are known are inconsistent with each other, so something has to be removed. We get a lot of letters from people insisting that we ought to makes holes in our guesses. You see, you make a hole, to make room for a new guess. Somebody says, 'You know, you people always say that space is continuous. How do you know when you get to a small enough dimension that there really are enough points in between, that it isn't just a lot of dots separated by little distances?' Or they say, 'You know those quantum mechanical amplitudes you told me about, they're so complicated and absurd, what makes you think those are right? Maybe they aren't right'. Such remarks are obvious and are perfectly clear to anybody who is working on this problem. It does not do any good to point this out. The problem is not only what might be wrong but what, precisely, might be substituted in place of it. In the case of the continuous space, suppose the precise proposition is that space really consists of a series of dots, and that the space between them does not mean anything, and that the dots are in a cubic array. Then we can prove immediately that this is wrong. It does not work. The problem is not just to say something might be wrong, but to replace it by something – and that is not so easy. As soon as any really definite idea is substituted it becomes almost immediately apparent that it does not work.

The second difficulty is that there is an infinite number of possibilities of these simple types. It is something like this. You are sitting working very hard, you have worked for a long time trying to open a safe. Then some Joe comes along who knows nothing about what you are doing, except that you are trying to open the safe. He says 'Why don't you try the combination 10:20:30?' Because you are busy, you have tried a lot of things, maybe you have already tried 10:20:30. Maybe you know already that the middle number is 32 not 20. Maybe you know as a matter of fact that it is a five digit combination. . . . So please do not send me any letters trying to tell me how the thing is going to work. I read them – I always read them to make sure that I have not already thought of what is suggested – but it takes too long to answer them, because they are usually in the class 'try 10:20:30'. As usual, nature's imagination far surpasses our own, as we have seen from the other theories which are subtle and deep. To get such a subtle and deep guess is not so easy. One must be really clever to guess, and it is not possible to do it blindly by machine.

I want to discuss now the art of guessing nature's laws. It is an art. How is it done? One way you might suggest is to look at history to see how the other guys did it. So we look at history.

We must start with Newton. He had a situation where he had incomplete knowledge, and he was able to guess the laws by putting together ideas which were all relatively close to experiment; there was not a great distance between the observations and the tests. That was the first way, but today it does not work so well.

The next guy who did something great was Maxwell, who obtained the laws of electricity and magnetism. What he did was this. He put together all the laws of electricity, due to Faraday and other people who came before him, and he looked at them and realized that they were mathematically inconsistent. In order to straighten it out he had to add one term to an equation. He did this by inventing for himself a model of idler wheels and gears and so on in space. He found what the new law was – but nobody paid much attention because they did not believe in the idler wheels. We do not believe in the idler wheels today, but the equations that he obtained were correct. So the logic may be wrong but the answer right.

In the case of relativity the discovery was completely different. There was an accumulation of paradoxes; the known laws gave inconsistent results. This was a new kind of thinking, a thinking in terms of discussing the possible symmetries of laws. It was especially difficult, because for the first time it was realized how long something like Newton's laws could seem right, and still ultimately be wrong. Also it was difficult to accept that ordinary ideas of time and space, which seemed so instinctive, could be wrong.

Quantum mechanics was discovered in two independent ways – which is a lesson. There again, and even more so, an enormous number of paradoxes were discovered experimentally, things that absolutely could not be explained in any way by what was known. It was not that the knowledge was incomplete, but that the knowledge was too complete. Your prediction was that this should happen – it did not. The two different routes were one by Schrödinger,* who guessed the equation, the other by Heisenberg, who argued that you must analyse what is measurable. These two different philosophical methods led to the same discovery in the end.

More recently, the discovery of the laws of the weak decay I spoke of, when a neutron disintegrates into a proton, an electron and an anti-neutrino – which are still only partly known – add up to a somewhat different situation. This time it was a case of incomplete knowledge, and only the equation was guessed. The special difficulty this time was that the experiments were all wrong. How can you guess the right answer if, when you calculate the result, it disagrees with experiment? You need courage to say the experiments must be wrong. I will explain where that courage comes from later.

Today we have no paradoxes – maybe. We have this infinity that comes in when we put all the laws together, but the people sweeping the dirt under the rug are so clever that one sometimes thinks this is not a serious paradox. Again, the fact that we have found all these particles does not tell us anything except that our knowledge is incomplete. I am sure that history does not repeat itself in physics, as you can tell from looking at the examples I have given. The reason is this. Any schemes – such as 'think of symmetry laws', or 'put the information in mathematical form', or 'guess equations' – are known to everybody now, and they are all tried all the time. When you are stuck, the answer cannot be one of these, because you will have tried these right away.

*Erwin Schrödinger, Austrian theoretical physicist. Won Nobel Prize for Physics 1933 with Paul Dirac.

There must be another way next time. Each time we get into this log-jam of too much trouble, too many problems, it is because the methods that we are using are just like the ones we have used before. The next scheme, the new discovery, is going to be made in a completely different way. So history does not help us much.

I should like to say a little about Heisenberg's idea that you should not talk about what you cannot measure, because many people talk about this idea without really understanding it. You can interpret this in the sense that the constructs or inventions that you make must be of such a kind that the consequences that you compute are comparable with experiment – that is, that you do not compute a consequence like 'a moo must be three goos', when nobody knows what a moo or a goo is. Obviously that is no good. But if the consequences can be compared to experiment, then that is all that is necessary. It does not matter that moos and goos cannot appear in the guess. You can have as much junk in the guess as you like, provided that the consequences can be compared with experiment. This is not always fully appreciated. People often complain of the unwarranted extension of the ideas of particles and paths, etc., into the atomic realm. Not so at all; there is nothing unwarranted about the extension. We must, and we should, and we always do, extend as far as we can beyond what we already know, beyond those ideas that we have already obtained. Dangerous? Yes. Uncertain? Yes. But it is the only way to make progress. Although it is uncertain, it is necessary to make science useful. Science is only useful if it tells you about some experiment that has not been done; it is no good if it only tells you what just went on. It is necessary to extend the ideas beyond where they have been tested. For example, in the law of gravitation, which was developed to understand the motion of planets, it would have been no use if Newton had simply said, 'I now understand the planets', and had not felt able to try to compare it with the earth's pull on the moon, and for later men to say 'Maybe what holds the galaxies together is gravitation'. We must try that. You could say, 'When you get to the size of the galaxies, since you know nothing about it, anything can happen'. I know, but there is no science in accepting this type of limitation. There is no ultimate understanding of the galaxies. On the other hand, if you assume that the entire behaviour is due only to known laws, this assumption is very limited and definite and easily broken by experiment. What we are looking for is just such hypotheses, very definite and easy to compare with experiment. The fact is that the way the galaxies behave so far does not seem to be against the proposition.

I can give you another example, even more interesting and important. Probably the most powerful single assumption that contributes most to the progress of biology is the assumption that everything animals do the atoms can do, that the things that are seen in the biological world are the results of the behaviour of physical and chemical phenomena, with no 'extra something'. You could always say, 'When you come to living things, anything can happen'. If you accept that you will never understand living things. It is very hard to believe that the wiggling of the tentacle of the octopus is nothing but some fooling around of atoms according to the known physical laws. But when it is investigated with this hypothesis one is able to make guesses quite accurately about how it works. In this way one makes great progress in understanding. So far the tentacle has not been cut off – it has not been found that this idea is wrong.

It is not unscientific to make a guess, although many people who are not in science think it is. Some years ago I had a conversation with a layman about flying saucers – because I am scientific I know all about flying saucers! I said 'I don't think there are flying saucers'. So my antagonist said, 'Is it impossible that there are flying saucers? Can you prove that it's impossible?' 'No', I said, 'I can't prove it's impossible. It's just very unlikely'. At that he said, 'You are very unscientific. If you can't prove it impossible then how can you say that it's unlikely?' But that is the way that *is* scientific. It is scientific only to say what is more likely and what less likely, and not to be proving all the time the possible and impossible. To define what I mean, I might have said to him, 'Listen, I mean that from my knowledge of the world that I see around me, I think that it is much more likely that the reports of flying saucers are the results of the known irrational characteristics of terrestrial intelligence than of the unknown rational efforts of extra-terrestrial intelligence'. It is just more likely, that is all. It is a good guess. And we always try to guess the most likely explanation, keeping in the back of the mind the fact that if it does not work we must discuss the other possibilities.

How can we guess what to keep and what to throw away? We have all these nice principles and known facts, but we are in some kind of trouble: either we get the infinities, or we do not get enough of a description – we are missing some parts. Sometimes that means that we have to throw away some idea; at least in the past it has always turned out that some deeply held idea had to be thrown away. The question is, what to throw away and what to keep. If you throw it all away that is going a little far, and then you have not much to work with. After all, the conservation of energy looks good, and it is nice, and I do not want to throw it away. To guess what to keep and what to throw away takes considerable skill. Actually it is probably merely a matter of luck, but it looks as if it takes considerable skill.

Probability amplitudes are very strange, and the first thing you think is that the strange new ideas are clearly cock-eyed. Yet everything that can be deduced from the ideas of the existence of quantum mechanical probability amplitudes, strange though they are, do work, throughout the long list of strange particles, one hundred per cent. Therefore I do not believe that when we find out the inner guts of the composition of the world we shall find these ideas are wrong. I think this part is right, but I am only guessing: I am telling you how I guess.

On the other hand, I believe that the theory that space is continuous is wrong, because we get these infinities and other difficulties, and we are left with questions on what determines the size of all the particles. I rather suspect that the simple ideas of geometry, extended down into infinitely small space, are wrong. Here, of course, I am only making a hole, and not telling you what to substitute. If I did, I should finish this lecture with a new law.

Some people have used the inconsistency of all the principles to say that there is only one possible consistent world, that if we put all the principles together, and calculate very exactly, we shall not only be able to deduce the principles, but we shall also discover that these are the only principles that could possibly exist if the thing is still to remain consistent. That seems to me a big order. I believe that sounds like wagging the dog by the tail. I believe that it has to be given that certain things exist – not all the 50-odd particles, but a few little things like electrons, etc. – and then with all the principles the great complexities that come out are probably a definite consequence. I do not think that you can get the whole thing from arguments about consistencies.

Another problem we have is the meaning of the partial symmetries. These symmetries, like the statement that neutrons and protons are nearly the same but are not the same for electricity, or the fact that the law of reflection symmetry is perfect except for one kind of reaction, are very annoying. The thing is almost symmetrical but not completely. Now two schools of thought exist. One will say that it is really simple, that they are really symmetrical but that there is a little complication which knocks it a bit cock-eyed. Then there is another school of thought, which has only one representative, myself, which says no, the thing may be complicated and become simple only through the complications. The Greeks believed that the orbits of the planets were circles. Actually they are ellipses. They are not quite symmetrical, but they are very close to circles. The question is, why are they very close to circles? Why are they nearly symmetrical? Because of a long complicated effect of tidal friction – a very complicated idea. It is possible that nature in her heart is completely unsymmetrical in these things, but in the complexities of reality it gets to look approximately as if it is symmetrical, and the ellipses look almost like circles. That is another possibility; but nobody knows, it is just guesswork.

Suppose you have two theories, A and B, which look completely different psychologically, with different ideas in them and so on, but that all the consequences that are computed from each are exactly the same, and both agree with experiment. The two theories, although they sound different at the beginning, have all consequences the same, which is usually easy to prove mathematically by showing that the logic from A and B will always give corresponding consequences. Suppose we have two such theories, how are we going to decide which one is right? There is no way by science, because they both agree with experiment to the same extent. So two theories, although they may have deeply different ideas behind them, may be mathematically identical, and then there is no scientific way to distinguish them.

However, for psychological reasons, in order to guess new thcorics, these two things may be very far from equivalent, because one gives a man different ideas from the other. By putting the theory in a certain kind of framework you get an idea of what to change. There will be something, for instance, in theory A that talks about something, and you will say, 'I'll change that idea in here'. But to find out what the corresponding thing is that you are going to change in B may be very complicated – it may not be a simple idea at all. In other words, although they are identical before they are changed, there are certain ways of changing one which looks natural which will not look natural in the other. Therefore psychologically we must keep all the theories in our heads, and every theoretical physicist who is any good knows six or seven different theoretical representations for exactly the same physics. He knows that they are all equivalent, and that nobody is ever going to be able to decide which one is right at that level, but he keeps them in his head, hoping that they will give him different ideas for guessing.

That reminds me of another point, that the philosophy or

ideas around a theory may change enormously when there are very tiny changes in the theory. For instance, Newton's ideas about space and time agreed with experiment very well, but in order to get the correct motion of the orbit of Mercury, which was a tiny, tiny difference, the difference in the character of the theory needed was enormous. The reason is that Newton's laws were so simple and so perfect, and they produced definite results. In order to get something that would produce a slightly different result it had to be completely different. In stating a new law you cannot make imperfections on a perfect thing; you have to have another perfect thing. So the differences in philosophical ideas between Newton's and Einstein's theories of gravitation are enormous.

What are these philosophies? They are really tricky ways to compute consequences quickly. A philosophy, which is sometimes called an understanding of the law, is simply a way that a person holds the laws in his mind in order to guess quickly at consequences. Some people have said, and it is true in cases like Maxwell's equations, 'Never mind the philosophy, never mind anything of this kind, just guess the equations. The problem is only to compute the answers so that they agree with experiment, and it is not necessary to have a philosophy, or argument, or words, about the equation'. That is good in the sense that if you only guess the equation you are not prejudicing yourself, and you will guess better. On the other hand, maybe the philosophy helps you to guess. It is very hard to say.

For those people who insist that the only thing that is important is that the theory agrees with experiment, I would like to imagine a discussion between a Mayan astronomer and his student. The Mayans were able to calculate with great precision predictions, for example, for eclipses and for the position of the moon in the sky, the position of Venus, etc. It was all done by arithmetic. They counted a certain number and subtracted some numbers, and so on. There was no discussion of what the moon was. There was no discussion even of the idea that it went around. They just calculated the time when there would be an eclipse, or when the moon would rise at the full, and so on. Suppose that a young man went to the astronomer and said, 'I have an idea. Maybe those things are going around, and there are balls of something like rocks out there, and we could calculate how they move in a completely different way from just calculating what time they appear in the sky'. 'Yes', says the astronomer, 'and how accurately can you predict eclipses?' He says, 'I haven't developed the thing very far yet'. Then says the astronomer, 'Well, we can calculate eclipses more accurately than you can with your model, so you must not pay any attention to your idea because obviously the mathematical scheme is better'. There is a very strong tendency, when someone comes up with an idea and says, 'Let's suppose that the world is this way', for people to say to him, 'What would you get for the answer to such and such a problem?' And he says, 'I haven't developed it far enough'. And they say, 'Well, we have already developed it much further, and we can get the answers very accurately'. So it is a problem whether or not to worry about philosophies behind ideas.

Another way of working, of course, is to guess new principles. In Einstein's theory of gravitation he guessed, on top of all the other principles, the principle that corresponded to the idea that the forces are always proportional to the masses. He guessed the principle that if you are in an accelerating car you cannot distinguish that from being in a gravitational field, and by adding that principle to all the other principles, he was able to deduce the correct laws of gravitation.

That outlines a number of possible ways of guessing. I would now like to come to some other points about the final result. First of all, when we are all finished, and we have a mathematical theory by which we can compute consequences, what can we do? It really is an amazing thing. In order to figure out what an atom is going to do in a given situation we make up rules with marks on paper, carry them into a machine which has switches that open and close in some complicated way, and the result will tell us what the atom is going to do! If the way that these switches open and close were some kind of model of the atom, if we thought that the atom had switches in it, then I would say that I understood more or less what is going on. I find it quite amazing that it is possible to predict what will happen by mathematics, which is simply following rules which really have nothing to do with what is going on in the original thing. The closing and opening of switches in a computer is quite different from what is happening in nature.

One of the most important things in this 'guess – compute consequences – compare with experiment' business is to know when you are right. It is possible to know when you are right way ahead of checking all the consequences. You can recognize truth by its beauty and simplicity. It is always easy when you have made a guess, and done two or three little calculations to make sure that it is not obviously wrong, to know that it is right. When you get it right, it is obvious that it is right – at least if you have any experience – because usually what happens is that more comes out than goes in. Your guess is, in fact, that something is very simple. If you cannot see immediately that it is wrong, and it is simpler than it was before, then it is right. The inexperienced, and crackpots, and people like that, make guesses that are simple, but you can immediately see that they are wrong, so that does not count. Others, the inexperienced students, make guesses that are very complicated, and it sort of looks as if it is all right, but I know it is not true because the truth always turns out to be simpler than you thought. What we need is imagination, but imagination in a terrible strait-jacket. We have to find a new view of the world that has to agree with everything that is known, but disagree in its predictions somewhere, otherwise it is not interesting. And in that disagreement it must agree with nature. If you can find any other view of the world which agrees over the entire range where things have already been observed, but disagrees somewhere else, you have made a great discovery. It is very nearly impossible, but not quite, to find any theory which agrees with experiments over the entire range in which all theories have been checked, and yet gives different consequences in some other range, even a theory whose different consequences do not turn out to agree with nature. A new idea is extremely difficult to think of. It takes a fantastic imagination.

What of the future of this adventure? What will happen ultimately? We are going along guessing the laws; how many laws are we going to have to guess? I do not know. Some of my colleagues say that this fundamental aspect of our science will go on; but I think there will certainly not be perpetual novelty, say for a thousand years. This thing cannot keep on going so that we are always going to discover more and more new laws. If we do, it will become boring that there are so many levels one underneath the other. It seems to me that what can happen in the future is either that all the laws become known – that is, if you had enough laws you could compute consequences and they would always agree with experiment, which would be the end of the line – or it may happen that the experiments get harder and harder to make, more and more expensive, so you get 99·9 per cent of the phenomena, but there is always some phenomenon which has just been discovered, which is very hard to measure, and which disagrees; and as soon as you have the explanation of that one there is always another one, and it gets slower and slower and more and more uninteresting. That is another way it may end. But I think it has to end in one way or another.

We are very lucky to live in an age in which we are still making discoveries. It is like the discovery of America – you only discover it once. The age in which we live is the age in which we are discovering the fundamental laws of nature, and that day will never come again. It is very exciting, it is marvellous, but this excitement will have to go. Of course in the future there will be other interests. There will be the interest of the connection of one level of phenomena to another – phenomena in biology and so on, or, if you are talking about exploration, exploring other planets, but there will not still be the same things that we are doing now.

Another thing that will happen is that ultimately, if it turns out that all is known, or it gets very dull, the vigorous philosophy and the careful attention to all these things that I have been talking about will gradually disappear. The philosophers who are always on the outside making stupid remarks will be able to close in, because we cannot push them away by saying, 'If you were right we would be able to guess all the rest of the laws', because when the laws are all there they will have an explanation for them. For instance, there are always explanations about why the world is three-dimensional. Well, there is only one world, and it is hard to tell if that explanation is right or not, so that if everything were known there would be some explanation about why those were the right laws. But that explanation would be in a frame that we cannot criticize by arguing that that type of reasoning will not permit us to go further. There will be a degeneration of ideas, just like the degeneration that great explorers feel is occurring when tourists begin moving in on a territory.

In this age people are experiencing a delight, the tremendous delight that you get when you guess how nature will work in a new situation never seen before. From experiments and information in a certain range you can guess what is going to happen in a region where no one has ever explored before. It is a little different from regular exploration in that there are enough clues on the land discovered to guess what the land that has not been discovered is going to look like. These guesses, incidentally, are often very different from what you have already seen – they take a lot of thought.

What is it about nature that lets this happen, that it is possible to guess from one part what the rest is going to do? That is an unscientific question: I do not know how to answer it, and therefore I am going to give an unscientific answer. I think it is because nature has a simplicity and therefore a great beauty.